T0298021

On Foundations of
SEISMOLOGY
Bringing Idealizations Down to Earth

On **Foundations** of
SEISMOLOGY
Bringing Idealizations Down to Earth

James Robert Brown

Michael A. Slawinski

With illustrations by Roberto Lauciello

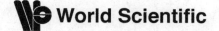

NEW JERSEY · LONDON · SINGAPORE · BEIJING · SHANGHAI · HONG KONG · TAIPEI · CHENNAI

Published by

World Scientific Publishing Co. Pte. Ltd.

5 Toh Tuck Link, Singapore 596224

USA office: 27 Warren Street, Suite 401-402, Hackensack, NJ 07601

UK office: 57 Shelton Street, Covent Garden, London WC2H 9HE

Library of Congress Cataloging-in-Publication Data
Names: Brown, James Robert. | Slawinski, M. A. (Michael A.), 1961–
Title: On foundations of seismology : bringing idealizations down to earth / James Robert Brown
    (University of Toronto, Canada), Michael Slawinski (Memorial University, Canada).
Description: New Jersey : World Scientific, 2017. |
    Includes bibliographical references and index.
Identifiers: LCCN 2017014048| ISBN 9789814329491 (hardcover : alk. paper) |
    ISBN 9814329495 (hardcover : alk. paper)
Subjects: LCSH: Seismology. | Continuum mechanics. | Earth sciences--Mathematics.
Classification: LCC QE534.3 .B76 2017 | DDC 551.22--dc23
LC record available at https://lccn.loc.gov/2017014048

British Library Cataloguing-in-Publication Data
A catalogue record for this book is available from the British Library.

Desk Editor: Ng Kah Fee

Typeset by Stallion Press
Email: enquiries@stallionpress.com

Printed in Singapore

photograph from Bradley's family collection

This book is dedicated to the late philosopher Jim Bradley and to those who strive to follow his spirit of respect for the principles of reason without fear of postulating hypotheses about the intrinsic, even if unattainable, nature of things.

Contents

List of Figures

Acknowledgments

Improvements of this book resulted from collaborations and discussions with Aitor Anduaga, Len Bos, Robert Batterman, David Dalton, Marcelo Epstein, Klaus Helbig, Misha Kochetov, Margo Kondratieva, Thomas Meehan, Michael Rochester, Yves Rogister, Sergey Sadov, Robert Sarracino, Theodore Stanoev and Albert Tarantola. Diagrams were created by Elena Patarini. Editorial work was done by David Dalton, Thomas Meehan, Elena Patarini and Theodore Stanoev. Kah-Fee Ng ensured the adjustment of final details prior to printing.

Preface

The reciprocal relation of epistemology and science is of a noteworthy kind.
They are dependent upon each other. Epistemology without contact with
science becomes an empty scheme. Science without epistemology is—in so
far as it is thinkable at all—primitive and muddled.

Albert Einstein (1949)

Early humans, to survive, needed to focus on their surroundings. Their
curiosity, however, extended to the heavens above and the Earth below.
What is below the Earth's surface? Are there people living in its interior?
Can it be a place where gods or demons dwell? Their questions were followed by speculative answers. After millennia, at least partial answers to
questions about the Earth's interior have come from seismology.

Seismology—by its methodology—is a branch of physics, and—by its
scope of applications—of the Earth sciences. In this book, beyond discussing its foundations, an examination of seismological theory gives us an
opportunity to discuss foundations of natural sciences, in general, since the
seismological theory is an example of a mediation between the observations
and the properties of the natural world. Herein, the observations are almost exclusively provided by seismographs, which detect disturbances that
propagate in the Earth.

The achievements of seismology over the past couple of centuries have
been significant. In this book, we investigate the approaches that led to
these achievements. However, it is neither a historical treatise of seismology
nor a textbook to study its techniques, but a discussion of its foundations,
in which we examine the conceptual structure of the theory.

Since foundational questions are philosophical, philosophers might find
this book interesting as an enquiry into idealizations, fictions and models,
whose use in seismology is extensive. Seismologists might find it insightful

in arranging these idealizations, fictions and models in a logical structure. Also, this book might be of interest to anyone who wishes to enquire into foundational issues of natural sciences, in particular, their epistemological and ontological questions.

Fig. 0.1

Fig. 0.2

The relation between philosophy and science is one of interaction. Philosophers consider that scientists who follow rational and logical procedures are justified in their conclusions. They would not say that astrologers are justified in their conclusions, as suggested by the comic strip presented in Figures 0.1 and 0.2. Among philosophical issues, there is the distinction between science and pseudoscience.

The relation between philosophy and science is symbiotic. Philosophy sheds light on conceptual complexities of seismological theory and seismology provides philosophy with rich examples and information about such philosophical topics as realism and idealization. Philosophers and scientists emphasize different aspects, but strive to reach a coherent understanding.

This book is a sample of foundational topics, which might be of interest, and even of practical value, to the readers. We address several issues that arise in scientific practices of seismology, but the emphasis is on the relation between the abstractness of continuum mechanics, which is the basis of seismological theory, and the physical interpretation of seismology.

Many topics are not addressed here. This does not imply that these topics are of lesser importance, but it is symptomatic of pragmatic limitations. For instance, the variational principles, in particular Fermat's and Hamilton's principles, are not discussed, even though they exhibit profound philosophical and scientific issues.[1,2] Also, the relation between seismic waves and rays[3]—which shares similarities with the relation between quantum mechanics and classical mechanics—is not examined, even though it invokes the interesting problem of asymptotic reasoning,[4] used in seismic ray theory.[5]

We ask the readers's indulgence on a couple of points. First, the book is written with a scope of the readership encompassing seismologists who have an interest in the foundations of their discipline, philosophers who have an interest in a physical science that is—so far—little studied by philosophy, as well as mathematicians, physicists and engineers who have an interest in continuum mechanics and its applications. Charitable readers should understand our need to make certain concessions in order to render this book accessible to a broad readership.

Second, we sharply distinguish among the three following concepts.

1. The *Earth*, whose mechanical properties are the ultimate target of our enquiry

---

[1] *see also*: Slawinski (2015, Part 3) and Slawinski (2016, Chapter 7)

[2] Readers interested in examining both the scientific and philosophical aspects of variational principles might refer to Kot (2014), Lanczos (1986/1970) and Hildebrandt and Tromba (1996).

[3] *see also*: Slawinski (2015, Part 2)

[4] Readers interested in mathematics of asymptotic methods and philosophical aspects of asymptotic reasoning might refer to Babich and Buldyrev (2009) and Batterman (2002), respectively.

[5] *see also*: Bóna and Slawinski (2015)

2. A model of the Earth, which we take to be a *continuum* of continuum mechanics
3. *Mathematics* that furnishes analogies to define and examine properties of that continuum, which allow us to make inferences of terrestrial properties

Examining foundations of seismology, we must distinguish these three concepts, but to do so at all times could become pedantic so, occasionally, we might follow common speech. Again, charitable readers should see this as shorthand for a longer but rather tedious account. Yet, as emphasized by the quote from Stebbing (1958/1937), we are aware that

> *exact* thought cannot be *conveyed in inexact language*; at best
> it can be but partially conveyed, at worst the illusion will be
> created that it has been conveyed.

This book is structured in the following manner. In Chapter 1, we provide an introduction to seismology. Also, we give an overview of issues to be discussed, such as scientific methods, theories and models.

In Chapter 2, we examine the concept of science and scientific methods. In other words, we address both ontological and epistemological issues.

In Chapter 3, we discuss primitive concepts of continuum mechanics, since it is the basic theory underlying seismology.

In Chapter 4, we present the theory of continuum mechanics, which—as a consequence of its primitive concepts—is divided into two parts: general principles and constitutive relations.

In Chapter 5, we focus our attention on the constitutive relation of a Hookean solid, whose mathematical properties serve as analogies for mechanical properties of the Earth.

In Chapter 6, we examine methods that allow seismologists to infer physical properties by comparing theoretical predictions with observations. We discuss it within the concepts of forward and inverse problems.

In Chapter 7, we discuss the intertheory and intratheory relations, such as reduction and emergence, on the one hand, and approximations and generalizations, on the other hand. In that context, we comment on similarities and differences between geology and geophysics.

In *Afterword*, we emphasize the essential points of our discussion and present, in Figure A.1, a concise illustration of relations among the Earth, seismology, continuum mechanics and mathematics.

The book contains many footnotes, which play an important—but secondary—part; herein, secondary means that the reading need not be interrupted by referring to any footnote, which can be consulted upon deeper study of the material. Since this book is the fourth volume of the series that contains Slawinski (2015), Bóna and Slawinski (2015) and Slawinski (2016), the footnotes marked by *see also* direct the reader for further details to specific locations in these volumes. Other footnotes refer to pertinent work whose reference might broaden the reader's perspective and facilitate the understanding of material.

A detailed subject index allows the reader to use this book as a reference. Therein, *see also* followed only by a term directs the reader to a synonymous or an analogous term; *see also* followed by a name and page numbers directs the reader to places in which the meaning of the initial term is clarified by a comparison with its opposition, generalization or reduction.

An extensive name index provides insight into the historical development of seismology. Therein, *see also* directs the reader to another, perhaps adopted, name for that person.

Chapter 1

Aspects of seismological theory

Given any rule, however "fundamental" or "necessary" for science, there are always circumstances when it is advisable not only to ignore the rule, but to adopt its opposite.

Paul K. Feyerabend (1975)

Preliminary remarks

Seismology is a quantitative physical science that examines mechanical properties of our planet. Descriptions of these properties result from an interplay of mathematical strategies, computational techniques and observational apparatus. Beside its intrinsic interest in describing the Earth's interior, seismology is of practical importance in earthquake predictions as well as mineral and petroleum exploration.

We begin this chapter with comments on seismology and its history. Then, we describe the conceptual issues involved in its theory, including the relation between theories and models, as well as the choices available to a theorist. We conclude this chapter with a discussion on scientific methodologies.

1.1 Seismology

Seismological theory underlies the study of the propagation of mechanical disturbances within materials composing the Earth and Earth-like bodies, such as planets and moons. There are also seismological studies of stars, referred to as asteroseismology, particularly of the Sun, which is referred to as helioseismology.

From the viewpoint of its inferences, seismology is a branch of geosciences. From the viewpoint of its methodology, seismology is a branch of physics, and—in particular—mechanics.

The term seismology was coined in the nineteenth century by one of its founders, Robert Mallet, in the context of the study of earthquakes; in Greek, $\sigma\varepsilon\iota\sigma\mu\acute{o}\zeta$ means earthquake and $\lambda o\gamma\acute{\iota}\alpha$ means study. Presently, the term is used in a generic sense, not unlike, say, nuclear physics, allowing for a variety of different theories. It does not imply a single and coherent structure as, say, general relativity.

A cluster of theories superseding each other forms the history of seismology. Let us mention a few of its highlights.

Since before the time of Aristotle, it has been known that the Earth is spherical. It was only natural to wonder about its interior and why there were such events as earthquakes. Religious explanations have been offered all along, declaring earthquakes to be expressions of God's anger, as illustrated in Figure 1.1. Perhaps surprisingly, such explanations are not only a thing of the past.

Fig. 1.1

Apart from supernatural explanations, there is also a long history of naturalistic accounts. One of these is that earthquakes are the product of a decaying Earth. Aristotle and his followers thought of the Earth as a living thing, a biological organism whose various parts had a function in sustaining the whole. Decay of the Earth could be part of the natural process.

Much thinking about the Earth's interior and about earthquakes is based on analogies with other well-known or newly discovered phenomena. Thus, in the eighteenth century, earthquakes were seen as similar to, and maybe caused by, electrical phenomena akin to thunder and lightning. Edmond Halley, in the late seventeenth century, proposed a hollow Earth. He suggested that it consisted of an 800-kilometer thick shell with two inner concentric shells that rotated in different directions. The reason for this concoction was to explain some anomalous magnetic phenomena. Leonhard Euler also proposed a hollow Earth with its own Sun at the centre and civilized people living on the inside.

Contemporary seismology is largely a product of research of the last two centuries. During that period, mathematical formulations for examining wave phenomena have been advanced and the quality and quantity of measuring instruments have been greatly augmented. Mathematical concepts of P wave, S wave, their surface, guided and interface counterparts, namely, the Rayleigh, Love and Stoneley waves (Love, 1944), as well as their anisotropic extensions, namely, qP, qS_1 and qS_2 waves, became part of the scientific terminology. Also, standards, such as the Richter (1935) scale,[1] developed in collaborations with Gutenberg, and the Preliminary Reference Earth Model (PREM), developed by Dziewoński and Anderson (1981), under the auspices of the Standard Earth Model Committee, were established.

There is, unfortunately, no recent comprehensive history of seismology, only earlier (Davison, 1927/2014) or brief (Dahlen and Tromp, 1998) accounts, as well as monographs and treatises with a particular historical, philosophical, social or cultural focus (Anduaga, 2016; Brush, 1980, 2009; Guidoboni and Poirier, 2004; Placanica, 1997).[2]

Seismology, in its current form, is a discipline of a quantitative nature, and hence, presents challenges in combining its mathematical and physical content. The rich variety of mathematical models and idealizations makes seismology an interesting theory to study. In light of these considerations, it is surprising that only a few have addressed the complex conceptual structure of seismology. Among them, the philosophers Aitor Anduaga, Mario Bunge and Sheldon Smith have contributed to a better understanding. Smith (2007b, p. 58) remarks that

---

[1] Readers interested in the life and work of Charles Francis Richter might refer to Hough (2007).
[2] *see also*: Slawinski (2016, Section 1.1)

> [...] the micro-structure that continuum mechanics attributes
> to bodies is not the structure that real-world bodies have. How-
> ever, one would never model real-world macroscopic bodies via
> quantum mechanics since that would involve an enormous *n*-
> body problem that is less likely to reveal the salient features of,
> say, actual billiard ball collisions than continuum mechanics.
> Thus, a modelling of billiard ball collisions that wants to take
> into account and predict the largest range of the features of
> actual billiard balls such as their deformation and constitutive
> responses including thermodynamic responses will tend to be
> formulated within classical continuum mechanics. Because of
> its robust contact with the actual world, this branch of physics
> can serve as a check on what is intuitively plausible when it
> comes to bodies.

Also, this robust contact makes an intuitive distinction between the
continuum-mechanics models and the actual world less obvious, which
might render the evaluation of seismological theory with its relation to
the Earth more difficult, an issue emphasized in this book.

Bunge (1998, pp. 568-571) points out the particular importance of seismology.

> A nice illustration of the intertwining of empirical and theoreti-
> cal events in the actual practice of science is offered by seismol-
> ogy, the study of elastic disturbances of Terra. [...] in order
> to read a seismogram so that it may become a set of data re-
> garding an event (e.g., an earthquake) or an evidence relevant
> to a theory (e.g., about the inner structure of our planet), the
> seismologist employs elasticity theory and all the theories that
> may enter the design and interpretation of the seismograph.

Again, the interplay between theory and observations is brought to our
attention, and—within observations—the theory of measuring devices is
also included.

With notable exceptions, there is little in the way of conceptual analysis
of the Earth sciences. Kleinhans *et al.* (2005) attempt to explain why it is
so.

> Earth science has received relatively little attention from
> philosophers of science. [...] most of Earth science is *Terra
> Incognita* to philosophers. Apparently, it is generally believed
> that Earth science cannot offer much excitement [...].

It turns out that intellectual excitement can indeed be found in seismology,
though perhaps not immediately. The conceptual problems of quantum

theory can be grasped early on, even if solutions seem out of reach. It is easy to be enticed by the uncertainty principle, by spooky action-at-a-distance, or by Schrödinger's cat that is both alive and dead, as illustrated in Figure 1.2.

Fig. 1.2

Yet, seismology's conceptual foundations are as challenging as those of quantum theory or general relativity, though less obvious. For instance, one has to engage in the study of continuum mechanics, which underlies seismology, before appreciating its conceptual richness.

Another reason for a philosophical lack of interest in seismology is its perceived dedication to mere practical applications. Charles Sanders Peirce (c.1896/1955, p. 53) understood this situation.

> Persons who know science chiefly by its results—that is to say, have no acquaintance with it as a living inquiry—are apt to acquire the notion that the universe is now entirely explained in its leading features; and that it is only here and there that the fabric of scientific knowledge betrays any rents.

Peirce is correct; seismology, in particular, seems far from the frontiers of science. All that remains is to use this theory to learn further details about the Earth. Peirce is also correct that this is an illusion.

Seismologists are far from completely understanding the Earth and perhaps even farther from understanding the theoretical foundations of their science. We find scientific practice of paramount interest and—in the spirit of Bunge (1998), Smith (2007b) and Peirce (c.1896/1955)—we see that seismology exhibits conceptual issues involving the intertwining of empirical, theoretical and mathematical concepts.

We formulate seismology by postulating four foundational principles.

    I. Earth is a granular body, intrinsically composed of discrete particles, but—for the purposes of seismology—assumed to be a physical continuum.

II. Continuum mechanics is the quantitative framework for seismology.

III. Constitutive equations of continuum mechanics specify properties of a material within a given model of the Earth.

IV. Contact between the constitutive equations and the Earth is made through seismic measurements whose interpretations are within the theory of seismology based on the quantitative framework of continuum mechanics.

In this book, as is common in continuum mechanics, the terms 'constitutive equation' and 'constitutive relation' are synonymous.

The justification for the four principles is provided by considerations such as that the wavelength of seismic disturbances is orders of magnitude larger than the size of grains within a medium through which they propagate. Hence, a continuum is an average of properties of granular materials over a wavelength. Thus, we can adopt the framework of continuum mechanics and bring to bear the formal apparatus of this theory. Observations need to be appropriately interpreted by the theory to provide the required evidence, where the theory of continuum mechanics mediates between observations and interpretations.

To learn about the Earth's interior by observing effects of earthquakes, seismologists study wave propagation. Measured results depend on properties of the materials through which waves propagate. However, deformations that travel below the surface of the Earth are not observable. One could ask if there exist disturbances that relate an earthquake to a measurement of vibrations at a distant seismograph. Different philosophical approaches to such a question are possible. An empiricist, for instance, might deny such disturbances, and be interested only in the correlations between observable events in one location with observable events in another.

1.2   Theories and models

In geosciences, one commonly learns about the two types of waves that propagate in the Earth, which are referred to as P and S. Strictly speaking, this is false.

P and S waves do not propagate in the Earth but are contained in the equation of motion within an abstract medium, which is a Hookean solid, used by seismologists to model the Earth. The existence of such waves in abstract media was formulated at the beginning of the nineteenth century by such mathematical physicists as Siméon Denis Poisson. Its applicability to interpret seismic measurements is credited to the work of the geologist Richard Dixon Oldham. In his quantitative interpretation, Oldham followed qualitative concepts of the eighteenth century geologist John Michell. As stated by Charles Davison (1927/2014, p. 23–24), in his historical treatise, Michell's permanent contribution to seismology was

> his distinction between the phenomena which are and are not essential to his theory. In separating the vibratory motion from the wave-like motion or visible waves, Michell was in advance of his time. He was one of the first, if not the very first, to assign the vibratory motion in earthquakes to the propagation of elastic waves in the earth's crust.

At first sight, emphasizing the difference between the Earth and its model may appear as a pedantic distinction, since it goes beyond ordinary scientific usage. However, in foundational studies it is important to understand the status of the entities in question. There are material entities, such as rocks, and there are abstract entities, such as numbers and rays, which are just as real, though they are neither material nor located in space and time. This is the realm of abstract entities, which holds not only mathematical objects but also models of the physical world. The distinction between the physical and abstract realms is referred to as Platonism, in honour of the ideas of Plato. The aim of a foundational study is partly for its intrinsic interest in the subject matter and partly to address conceptual problems that arise in the theory and which, in turn, give rise to misunderstandings and impede progress.

A medium commonly used in seismology is a Hookean solid, named in honour of Robert Hooke,[3] an English physicist of the seventeenth century. The equation of motion within an isotropic Hookean solid contains, as mentioned above, two wave equations that describe the propagation of the P and S waves. These waves are good analogies of observed disturbances, even though they exist in Hookean solids, not in the Earth.

As discussed in Chapter 5, below, the choice of a Hookean solid is a tradeoff between accuracy and simplicity. If seismologists choose a more complicated model—which might be more appropriate in certain cases—

---

[3] Readers interested in life and achievements of Robert Hooke might refer to Chapman (2005).

and if their observational methods are sensitive enough to detect subtle disturbances, they can associate these disturbances with other waves contained in the corresponding equation of motion within a more complicated model.

Measurement presents interesting problems. Herein, by measurement we do not mean anything akin to the quantum-theory measurement, which might involve the creation of measured properties by the very act of measurement. We take the common-sense view that there are wavelike disturbances in the Earth, whether seismologists measure them or not. Our issue is different but equally intriguing.

The model seismologists choose determines the measurements they make. In choosing an isotropic Hookean model for the Earth, seismologists measure P and S waves as disturbances propagating through the interior of the Earth, and consider other measured disturbances as noise.

This is an instance of the so-called theory-ladenness of observation, a topic introduced in Section 1.3.3, below. In general, by invoking the theory of continuum mechanics, seismologists seek to observe entities, such as P and S waves, that do not exist without postulating that theory. We do not mean that the theory creates these phenomena, only that we need the concepts such as P and S waves in order to observe them.

Our knowledge of the Earth's interior necessarily relies on theoretical models. After all, we have not explored the interior directly, as anticipated by Jules Verne in his novel *Journey to the Centre of the Earth*, and illustrated in Figure 1.3.

Moreover, such a direct exploration would not suffice to answer many of the seismological questions, since it might be impossible to infer the response of the Earth as a whole by studying properties of small samples within the Earth.

Seismology is similar to other theories of natural science, such as electromagnetism and astrophysics, in having a hypotheticodeductive formulation. In calling seismological theory hypotheticodeductive, we are emphasizing its conjectural nature, as opposed to theories that are descriptive. Unlike the more descriptive sciences, such as palaeontology or entomology, which collect and classify specimens, seismology is conjectural; it postulates principles and derives their consequences. This is not a sharp distinction, since the taxonomic sciences inevitably theorize and the theoretical sciences inevitably collect and classify empirical data, but seismology is more on the theoretical and conjectural side. Among hypotheticodeductive theories, the observable consequences play a crucial role in evaluating the theory. The conjectured theory must predict or explain observations—and it must

Fig. 1.3

take false predictions to be counterexamples that refute the theory. It also matters that certain observations be novel, that is, unexpected except in the light of the theory. Predicting that the Sun will rise tomorrow is no achievement, but correctly predicting an earthquake well in advance would be, even though such predictions would be only probabilistic.

Other aspects also matter in theory assessment. Is a given theory consistent with other theories? Does it explain, and hence, unify various phenomena previously thought distinct? Is it simple, within concepts of simplicity discussed in Section 1.3.2, below? Is it elegant, which means that it contains substantial information expressed in a concise manner? These are important features, and philosophers of science debate the details and the relative importance of each.

Unlike quantum mechanics, seismology does not attempt to investigate the structure of matter. It postulates a continuum, similar to the electromagnetic field, but unlike Maxwell's electromagnetism, seismology cannot stand alone—it is based on continuum mechanics.[4]

---

[4]Readers interested in similarities and differences between continuum mechanics and electromagnetism might refer to Bunge (1967, Chapters 3.2 and 4.1).

Like other theories, seismology distinguishes theoretical entities and processes from the observational ones. This distinction, though commonplace, is not straightforward, and we discuss it in Chapter 2. Compared to other scientific theories, seismology is unusual in the extent to which it uses idealizations relating its mathematical and empirical aspects.

Among seismologists, it is common to refer to their pursuit as both an art and a science (Landa and Treitel, 2016). This sentiment stems from the *ad hoc* use of various idealizations and their interpretations. Referring to the interwar period, Anduaga (2016, p. 156) writes

> In academic earthquake seismology, as in commercial exploration, seismic interpretation was a subjective process [...] based on the *subjective* recognition of arrivals of energy or phases—*nota bene*, phases which the seismologist believed were worth plotting and selecting.

However, the fact that seismology invokes many idealizations and is burdened with conceptual uncertainties of interpretation does not make it unscientific.

1.3   Choice of models

1.3.1   *Underdetermination of theory by evidence*

In view of the statement of Box and Draper (1987), that, essentially, all models are wrong but some are useful, let us examine the choice of models in seismology. Herein, by 'wrong', we mean that no model, which belongs to an abstract realm, is tantamount to its physical counterpart; by useful, we mean that—in spite of being wrong—some models correctly predict and systematize empirical data.

Seismology provides a systematic framework for describing and predicting properties of the Earth. As for any inductive process, its conclusions extend beyond the observable evidence. There are many distinct models that are compatible with the available seismic data, and the choice of any model reaches beyond its immediate evidence. This is not unusual; it is typical of the sciences, in general. Furthermore, the epistemic problem of determining which, if any, model is correct, cannot be settled on the basis of observations alone. The problem of underdetermination of theory by evidence, as it is called, is the problem of having insufficient evidence to choose the correct theory. The notions of model and underdetermination are discussed in Chapter 2; herein, we offer only a few words of explanation.

The strong version of underdetermination is concerned with the claim that—for any set of data—there are infinitely many theories, even if not yet formulated, that can account for that set. The weak version concerns the problem that for a given set of data, we cannot choose among proposed rival theories, though future data might settle the issue. The strong version is a philosophical problem. The weak one is both a philosophical and a practical problem. It can be illustrated by the following example.

Example 1.1. Ptolemy and Copernicus

At the time of Nicolaus Copernicus, when his heliocentric and Claudius Ptolemy's geocentric theories made the same empirical claims, observations arguably could not decide between them. They made the same predictions about the observed location of Mars, the occurrence of an eclipse, and so on. It was only several centuries later that stellar parallax decided the issue empirically in favour of Copernicus.

Herein, we are concerned with the weak version of underdetermination. It is philosophically interesting and pragmatically important, since it is concerned with actual scientific practice. In seismology, weak underdetermination arises in many guises. Consider three inferences.

The first inference addresses mechanical properties of the outer core of the Earth, the second addresses the size of the core, and the third refers to the existence of the inner core.

Example 1.2. Properties of Earth's outer core

To consider mechanical properties described by seismological theory, it is common to model the Earth by concentric layers composed of Hookean solids, as illustrated in Figure 1.4.[5] According to such a model, mechanical properties change with depth. Furthermore, it is common to assume initially that the Hookean solid is isotropic, which, as already noted, implies that there are two types of waves within it: P and S. Since the properties of the Hookean solid change with depth, so do the velocities of these two waves.

Following a Hookean model, seismic observations indicate that speeds of P and S waves tend to increase with depth. However, at the boundary

---

[5]Readers interested in further details of the radial Earth model might refer to Kennett and Bunge (2008, Figure 1.6).

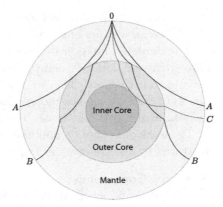

Fig. 1.4   A model of the Earth consisting of the mantle, the outer and inner cores. A source of disturbances, located at 0, results in their propagation illustrated by seismic rays.

between the mantle and the outer core, the P-wave speed drops to a lower value and the S-wave speed drops to zero, which means that S waves do not propagate in the outer core. This entails that the rigidity parameter of the model, contained in the equation of motion in which it describes the resistance to changes of shape, is zero.

P waves can be viewed as a result of the solid's resistance to changes of volume and shape, and S waves as a result of its resistance to changes of shape only. Thus, the material within the outer core resists a change of volume by responding with propagation of P waves, but not a change of shape, since it does not respond with propagation of S waves.

A resistance to a change of volume but not a change of shape is a characteristic property of a liquid. Even though we might not be able to put in more liquid than a given container allows, we can put it in containers of the same volume but different shapes.

The elegance of the seismological argument prompted a common, yet mistaken, view that the existence of the liquid outer core was inferred through seismology, namely, from the nonpropagation of S waves through that region of the Earth. However, the inference of the liquid outer core resulted from the work of Sir Harold Jeffreys on the Earth's tides. Notably, he also contributed to the theory of scientific inference (Jeffreys, 1931).

Seismic reflections were used to detect the boundary between the mantle and the core, which implies a change of mechanical properties but not a change from solid to liquid. If the outer core is postulated to be a liquid, no S wave can be transmitted through it. This results in the so-called S-wave shadow zone, spanning between the points denoted by A, in Figure 1.4.

Textbooks commonly claim that the existence of the liquid outer core was inferred in that manner. However, the lack of detection of S waves would not have been sufficient to draw this conclusion, particularly, if we consider the accuracy of measurements and the sparseness of data at that time. The evidence for a liquid core came from considerations raised by Jeffreys. He established that the rigidity of the mantle is much higher than the average rigidity of the entire Earth, from which he concluded that the core must have very low rigidity; hence it must be a liquid. A more detailed discussion of the discovery can be found in Brush (1980), Rochester and Crossley (1987) and Smith (2007a). Not only does Brush (1980) clarify the historical record, but he also points out mistaken lines of reasoning that persist to this day.

To explain the argument that leads to the false claim, let us emphasize that the concept of sharp shadow zones is based on simplifying assumptions. First, it is assumed that only transmitted waves are recorded or examined; in other words, no reflected waves—in particular, no waves reflected from the inner core—are taken into consideration. Second, no converted waves—in particular, no waves travelling as P in the outer core and as S in the inner core and the mantle—are considered. Another assumption, which results in the same trajectories for both P and S waves within the mantle, drawn in Figure 1.4 as rays from 0 to A, is the constancy of the ratio of their speeds within the mantle, which is a simplification of the Preliminary Reference Earth Model (PREM), mentioned in Section 1.1, above. It is a simplification, since the ratio of the two speeds varies depending on material properties of a Hookean solid. Notably, that ratio carries information about these properties, and as such is used in seismological studies to obtain Poisson's ratio,[6] which in this model is assumed to be constant within the entire mantle.

Remark 1.1. Using an isotropic Hookean solid to infer the existence of a liquid does not lead to inferring properties of such a liquid. Such an inference is impossible, since a Hookean solid is not a model of a liquid.[7] Setting the rigidity parameter of a Hookean solid to zero implies a solid with no resistance to change in shape. Let us also mention that, rigourously,

---

[6] *see also*: Slawinski (2015, pp. 194–197, pp. 199–201) and Slawinski (2016, p. 206)
[7] *see also*: Slawinski (2015, Section 3.2)

this parameter only tends to zero to satisfy the stability conditions for the Hookean solid, which are introduced in Section 5.3.2.[8]

As a further example of seismological reasoning, let us infer the size of the Earth's core.

Example 1.3. Size of Earth's core
Remaining within the spherical model of a Hookean solid, using seismological theory, and given a point on the surface of the Earth from which rays emanate, we can draw these rays, as illustrated in Figure 1.4. Once the initial angle is chosen, the ray is determined. Rays turn away from the centre of the Earth, if material properties are such that the speed of signal propagation increases with depth, and turn towards it, if the speed decreases. Since the speed is mostly increasing with depth, most signals emerge without reaching the centre. Only the signal propagating directly towards the centre reaches it. The presence of the liquid outer core results in the P-wave and S-wave shadow zones, which extend, respectively, between the points denoted by A and B and both points denoted by A.

For the P waves, the shadow zone is due to their decrease of propagation speed at the boundary between the mantle and the outer core, which entails an abrupt bending of the ray towards the centre. For the S waves, it is due to their incapacity to propagate in a liquid.

Given the laws of propagation and the model we adopt, these zones allow seismologists to infer the size of the core, since—given an earthquake near the surface—the P and S waves that are neither reflected nor converted are not detectable in the corresponding shadow zones.

The bending of the ray is a function of the signal-propagation speed, which is implied by the model. Modifying properties of the model results in changes of the bending. Consequently, what we take to be the size of the outer core depends on the model. Just as theories are underdetermined by evidence, so are models. There is a limited number of recording stations to delineate the shadow zones on the surface of the Earth. The measurements do not fully constrain the model. Yet, seismologists are comfortable with their predictions of the existence, mechanical properties, and size of the outer core.

Let us examine the third inference that exemplifies weak underdetermination in seismology. As the term outer core suggests, there is also an inner core. Its existence and properties were inferred by Inge Lehmann in the

---

[8] *see also*: Slawinski (2015, Section 4.3 and Section 5.12.4)

first half of the twentieth century by means of an analysis of subtle measurements in the context of a Hookean model to retrodict waves reflected
from the inner-core–outer-core boundary (Brush, 1980).

Example 1.4. Existence of Earth's inner core

In considering the shadow zones in Example 1.2, only the strongest effects
of propagating disturbances are modelled. Consequently, in Figure 1.4, the
P-wave shadow zone extends from A to B and the S-wave shadow zone
extends between both points marked by A. There is no allowance for any
effect of the inner core on these zones. If we consider reflections from the
inner core, we obtain P-wave arrivals within shadow zones, as illustrated in
Figure 1.4 by the ray whose endpoint is C. Thus—if we allow for reflections
within the model—there is no P-wave shadow zone.

Such considerations, which allow for modelling less strong effects of
disturbances, led Lehmann to interpret detected P waves as resulting from
the presence of the inner core. The boundary between the inner and outer
cores is called the Lehmann discontinuity.

Removing the assumption of no converted waves results in no S-wave
shadow zone. For instance, the ray connecting the source with the antipodal
point can correspond to a P wave or to an S-wave–P-wave combination.
However, converted waves describe the effects of even weaker disturbances
than the reflected ones.

Comparing the outer-core and inner-core examinations, we see that the
former requires that we restrict our discussion to fewer phenomena by ex-
cluding reflected or converted waves. The latter, on the other hand, requires
that we include these very phenomena. By ignoring reflected waves we are
able to recognize the existence of the outer core, infer its size and its liquid
nature. It is not uncommon to learn specific information about reality by
means of a simpler model. Persisting at this level of investigation, however,
we remain ignorant of the inner core. Its existence and size are revealed
only by including reflected waves.

Furthermore, the discussed outer-core and inner-core examinations implicitly assume the following idealization. Invoking the very concept of a
ray, seismologists assume that all the energy of the signal travels along
the ray in accordance with Fermat's principle of stationary time, according to which the transmitted signal travels along the least-time trajectory.[9]
This assumption ignores, among other things, the contribution of signals

---

[9] *see also*: Slawinski (2015, Section 13.1)

travelling through other paths, which might be significant. Seismologists—motivated by the concept of constructive interference[10]—ignore signals that do not follow the least-time trajectory by assuming that they are significantly weaker than signals propagating along the rays. Thus, the concept of shadow zones requires constraining of the energy to the ray itself. A more general theory might not allow seismologists to confirm physical properties of the Earth.

Shadow zones are not areas of a physical lack of disturbance, but theoretical areas where transmitted signals travelling along particular rays are absent. An observer unaware of the theory would not recognize the shadow zones in seismic records.

1.3.2 *Simplicity*

There are several other issues that deserve comment concerning the mode of inference used by seismologists. For instance, the Hookean model and observations are compatible with the existence of a liquid outer core. Yet, other models might not lead to this conclusion. Seismology interprets the data from seismographs as evidence for a particular theory of the structure of the Earth's interior. Shadow zones allowed us to infer the size of the core, while the Earth's rigidity was the first source of evidence for its existence.

Though our focus is on seismology, it must work in harmony with other geophysical approaches to the Earth. Paleomagnetism, for instance, might provide information in the context of plate tectonics to support inferences of seismological theory. The aim is an empirically adequate and systematic theory of the entire Earth.

Since seismologists have no direct access to the interior of the Earth, they can only model it by making various assumptions. Why choose a given model if other models are possible, in the sense of being compatible with observations? There is no decisive answer, just a number of plausibility considerations.

A Hookean solid is defined by a linear relation between two theoretical quantities that are at the essence of continuum mechanics: the stress and strain tensors. This first-order relation models deformations caused by mechanical disturbances. Model predictions agree—with a satisfactory accuracy—with experiments, provided deformations are small.[11] However,

---

[10] *see also*: Slawinski (2016, Section 7.2.6.1)
[11] Readers interested in isotropic models of behaviour for linear elasticity might refer to Fowler (1997, Section 7.2).

since a Hookean model is an idealization—which, beyond higher-order mechanical effects, ignores effects of temperature changes and heat transfer[12] due to wave propagation—it cannot be in complete agreement with physical observations.

It might be thought that these choices stem from Occam's law of ontological parsimony, known commonly as Occam's razor. However, this is not exactly the case. Occam's razor tells us to choose the explanation with the fewest assumptions, which is more likely to be true than explanations requiring more assumptions. His principle refers to physical truth. Herein, however, we are dealing with idealizations, and knowingly make assumptions that are false. There is an overlap in spirit with Occam's razor, but they are distinct concepts.

It is important to distinguish between two types of simplicity. Ontological simplicity refers to the number and kind of objects and processes postulated by a theory. Epistemological simplicity refers to the relative ease in using a theory in various applications.[13] One might claim that Ptolemy's Earth-centred theory is ontologically less simple than Copernicus's theory, since Ptolemy postulated distinctions among the Earth, Sun, and other planets. Copernicus postulated only the Sun and planets, with the Earth being one of the planets. In terms of practical simplicity, things could be reversed. Navigators find it easier to use Ptolemy's theory, where their position on the Earth is the centre of a coordinate system.

A Hookean solid is an ontologically simple model for the Earth. Other models, such as the Kelvin-Voigt model, discussed in Section 7.3, might involve other mechanisms. In certain aspects, a Hookean solid is also simple to use; notably—due to linearity—it allows for the use of the superposition principle: the final deformation is the sum of individual deformations, independently of their sequence.

Scientists and philosophers often elevate simplicity to an æsthetic and metaphysical status. They postulate simplicity of the physical world, and consider simplicity as a sign of truth in the ontological sense. However, neither type of simplicity should excessively hinder empirical accuracy, nor should accuracy be the sole criterion in constructing a theory.

---

[12] *see also*: Slawinski (2016, Section 8.3)

[13] Readers interested in an epistemological simplicity of seismology in the context of Bayesian statistics,[14] as well as its alteration within the structure of the petroleum industry, might refer to Anduaga (2016, Chapter 4).

[14] *see also*: Slawinski (2016, Section 9.3)

1.3.3   *Theory-ladenness of observation*

Once we have chosen a model of the Earth, certain types of observation are possible, but others are not. The theory-ladenness of observation is the doctrine that theories and background beliefs condition what we see, as further discussed in Section 2.1.4.3, below. Gestalt figures, such as the famous duck-rabbit shown in Figure 2.2 on page 37, illustrate this concept. The figure that impinges on the retina is the same for all observers, but one person might see a duck while another sees a rabbit. This phenomenon shows that we do not make neutral observations, but see things in the light of a theory. In the social sciences and in daily life, this can be dramatic. In our case, it has the consequence that seismologists observe P and S waves, but no other waves. Other things we could detect are passed off as noise. If seismologists used a different model for the Earth, other waves could be observed.

The theory-ladenness of observation does not undermine scientific objectivity, though it does complicate it. Some commentators claim that it implies that a theory's observable predictions are self-fulfilling. This is not so. On the assumption of a Hookean solid, seismologists might predict both P and S waves, whereas measurements could show only P waves, thus refuting a specific theory. Theory-ladenness only means that one observes properties and processes as they are conceptualized by the theory. It does not imply that the magnitudes of those properties will be as the theory predicts. Thus, there is plenty of scope for falsification of a model or even a theory.

1.4   Methodology

1.4.1   *Inference to best explanation*

The formulation of a model used by seismologists results, at least in part, from considerations drawn from seismology and from related— noncompeting and conciliatory—theories.

Newtonian gravitation competes with general relativity, but neither of them competes with the theory of evolution. Two theories are conciliatory if they imply similar consequences, thus reinforcing one another.

The existence of the liquid outer core was inferred by Jeffreys from the rigidity of the Earth, based on information about forced oscillations, which are the Earth's tides. These oscillations involve the deformation of the entire planet, similar to the vibrations of a bell ringing in a belfry.

The inference of the outer core is supported by the agreement between distinct approaches of seismology and global geodynamics, which increases confidence in the conclusion. Such a mode of inference is akin to what is known as Inference to the Best Explanation (IBE) (Lipton, 2004). Needless to say, there are debates over criteria of what constitutes the best explanation. There are also other serious issues concerning IBE, since many philosophers object, in principle, to this style of inference.

1.4.2 *Falsificationism*

Let us address the manner in which seismologists adjust the model to render it empirically adequate. It is possible to adjust values of many parameters that define a Hookean solid, since they are not measured. Seismologists conjecture these values and deduce observables to be compared with measurements. The values are accepted if there is a satisfactory agreement; otherwise, they are adjusted. The process continues until a satisfactory set of parameters is found.

Karl Popper (1959, 1963), whose approach to philosophy of science is discussed further in Section 2.1.3, advocated a doctrine of falsificationism, meaning that a theory is scientific only if it makes bold conjectures and could—at least, in principle—be refuted by a false prediction. If a theory does not make refutable-in-principle predictions, then it is not legitimate science. If it can be adjusted continuously to fit the data, it is no better than a pseudoscience.

According to Popper, seismology would seem to violate the canons of good science, being a cluster of assumptions and hypotheses that can be constantly revised without ever abandoning the theory. Seismologists might assume initially that the model for the Earth is spherically symmetric, that the Earth can be modelled by a mechanical continuum, and that the postulated continuum can be a Hookean solid described by elasticity parameters to be determined by observations. If a prediction does not agree with observations, one can ask the following question. What has been falsified? It may be hasty to reject the entire theory. The initial adjustment might be accomplished by modifying the values of elasticity parameters, which, however, might be too malleable.

This is contrary to Popper's doctrine. Yet, there are historical precedents of such a manner proceeding with fruitful results, as illustrated by the following example.

Example 1.5. Newton's lunar theory

Newton received observational data concerning the Moon's position from the Astronomer Royal, John Flamsteed. He wanted to use this data set to calculate the orbit of the Moon.

However, Flamsteed's data did not agree with Newton's lunar theory. In response, Newton rejected the data, not his own theory. He claimed that Flamsteed had used an inappropriate expression for the index of refraction when calculating the Moon's position. A different expression—and, hence, a different value—was used and yielded results that were in accord with Newton's theory. Newton had no independent measure of the index of refraction, but chose to set the value appropriate for his theory, and his choice was confirmed by further studies.

Even though this may seem arbitrary, it is not. For instance, there is no index of refraction that would save Ptolemy's theory. The range within which parameters can be adjusted is limited to plausible values. On the other hand, it is true that sufficiently complicated models might allow us to account for data by contradictory scenarios.

In view of practical considerations, such as availability of data, it might be unreasonable to argue for a complicated model. However, if the motivation for simplicity is the availability of data or ease of mathematical formulations, this motivation is independent from claims about reality. In view of the paucity of data, seismologists choose to remain within a simple model that is consistent with inferences of planetary physics, such as wobbles described by geodynamics, as well as Earth's formation and rotation. In other words, the background model is—in part—based on simplicity considerations, but it is also—in part—based on the prior and independent knowledge of other theories.

Closing remarks

Terms such as theory, model and idealization are common in science, but they need to be examined in a work on the foundations of the subject. In particular, such an examination is important within seismological theory, in which models and idealizations play an important role.

In this book, we introduce problems of idealization, including the construction of models and their testing, the relation between these models and the mathematical realm, and the relation between these models and the physical world. Most foundational questions do not admit final answers. However, by grappling with these issues, we develop critical awareness of, and learn about, the reliability and limits of seismological theory.

Chapter 2

Nature of science and its methods

La science, mon garçon, est faite d'erreurs, mais d'erreurs qu'il est bon de commettre, car elles mènent peu à peu à la vérité.[1]

Jules Verne (1864)

Preliminary remarks

Commonly, the term scientific method is used with approval and authority not only by scientists, but also by the general public. If an idea is acquired by the scientific method, then, according to the general consensus, we can trust it. Thus, people trying to convince us that astrology, creationism and homeopathy are legitimate sciences imply the use of scientific methods.

Scientific method requires reasoning and—depending on the approach and the nature of a given branch of science—is associated, to different degrees, with observations. There is, however, little agreement in proposing a general definition.

[2]We begin this chapter with a discussion of several epistemological issues. Subsequently, we address ontological issues. Both are central to the philosophy of science; the former refers to the scientific method and nature of evidence; the latter raises issues about reality and the aim of science.

---

[1]Science, my lad, is made up of mistakes, but they are mistakes that it is good to make, because they lead little by little to the truth.

[2]*see also*: Slawinski (2016, Chapter 9)

2.1   Epistemological issues

2.1.1   *On scientific methods*

Typical questions, which are central in choosing an appropriate methodology, are as follows. How are our beliefs justified? When should they
be overthrown? Is there a fundamental difference between beliefs based
directly on observation and those concerning theoretical entities? Is observation itself conditioned by our theoretical predictions or other beliefs?
What role, if any, do values play in the formation of our beliefs?

There is a wide range of methods, both general and specific, to address
such questions. Chemists use litmus paper to distinguish an acid from a
base, a technique of no concern to, say, a cultural anthropologist. Randomized clinical trials, introduced to cope with the influence of expectations,
are an appropriate method for testing the effectiveness of a drug, but they
are of no particular use in geophysics, since rocks are not subject to the
placebo effect.

The use of litmus paper and randomized trials are effective—but
specific—methods. The focus of this chapter, instead, is at the level of
general principles that guide us in science.[3]

2.1.2   *Positivism and empiricism*

2.1.2.1   *Logical Positivists*

A variety of guiding principle have been proposed for scientific pursuits.
The school of thought referred to as Logical Positivism was composed of
philosophers, mathematicians and scientists associated within the Austrian
capital; hence, they are commonly referred to as the Vienna Circle. For
many years, this group, which included the philosophers Moritz Schlick,
Rudolph Carnap and Otto Neurath, the physicist Philipp Frank, and the
mathematicians Karl Menger, Hans Hahn and Kurt Gödel, met regularly
until it was disbanded when the Nazis came to power.

Bertrand Russell's logic had a great influence on the approach of the
group, hence the term logical in their name.[4] Also, the influence of Ernst
Mach and Albert Einstein was of great importance. Mach, the philosopher-

---

[3]Readers interested in historical development of scientific methods might refer to original papers collected in the anthologies edited by Curd and Cover (1998) and by Boyd
et al. (1991).

[4]Readers interested in foundational issues of that period might refer to Doxiadis and
Papadimitriou (2009).

scientist, was the spiritual father of the Circle. Einstein, whose special relativity appeared based entirely on measurements made with rods and clocks—that is, without any appeal to hypothetical, nonobservable entities—appealed to the Positivist outlook.

Similar attitudes could be found elsewhere. American pragmatism, developed by Charles Sanders Peirce, William James and John Dewey, and the operationalism of Percy William Bridgman share many concepts with Logical Positivists. In psychology, behaviourism is motivated by the same philosophical ideas.

The main idea includes two related items: empiricism and verificationism. Empiricism assumes that all knowledge is based on sensory experiences. According to verificationism, for a scientific statement to be meaningful, it must be empirically testable; otherwise, the statement is neither true nor false; it is meaningless.

2.1.2.2 *Observational and theoretical terms*

Empiricism's historical rival is rationalism, a doctrine that accepts that some of our knowledge has a nonsensory source. Mathematics lends itself to such an interpretation, since perfect circles, imaginary numbers, and infinite sets are not the stuff of sensory experience. Ethics, too, is problematic for an empiricist, since observation can tell us who killed whom, but cannot tell us that such a killing is morally wrong. We need rational considerations to make that judgement. Within the natural sciences, common sense seems to support empiricism against the nonsensory sources of knowledge. Everyday examples support such a view: 'snow is cold' and 'rocks are hard' are straightforward examples of empirical knowledge. However, can empiricism do justice to all of science? No one can see electrons, magnetic fields, entropy, seismic rays or Freudian superegos. Thus, can we make sense of such theoretical concepts?

The Positivist approach to this challenge is to split the language of science into two distinct vocabularies, based on the apparent fact that some of the things we talk about are observable and others are not. Observational terms would include such expressions as apple, red, Earth. Theoretical terms would include electron, Hookean solid, P wave. The meaning of an observational term is acquired directly by a relevant experience. The only way to give meaning to theoretical terms is to link them—in an appropriate manner—to observational terms.

At first sight, the distinction between observational and theoretical terms—though important in such fields as particle physics—might appear beside the point in seismology. However, let us consider the following sentence. The outer core is liquid. Even though this sentence is composed of observational terms, its truth is not obtained by a mere observation. Determining the structure of the Earth's interior was not unlike Ernest Rutherford's firing alpha particles to determine the inner structure of the atom. The difference between a positively charged nucleus being a theoretical statement and the liquid outer core being hypothetically observable is minor. After all, even though liquid and outer core may be observational terms, the above sentence is not an observational statement. It is theoretical, since—at least at present—it can be tested only in an indirect manner.

Two morals could be drawn from this example. The Positivist assumption that the distinction between the theoretical and observational terms carries over to the distinction between resulting sentences is false. Also, language itself is not as intrinsic to understanding and to guiding the working of science as believed by Positivists. They had mistakenly thought that understanding science is tantamount to understanding scientific language.

There are further subtleties. By postulating the equations of motion in an isotropic Hookean solid, we deduce two types of waves: P and S; these waves are contained within equations in that solid.[5,6] An observation of disturbances received along a sequence of sensors on the surface of the Earth is interpreted as the physical analogue of these two waves. However, an alternative model, say, based on anisotropic Hookean solids, results in three types of waves, and hence, the same observation is interpreted as the analogue of three waves. In the mathematical model, the S wave splits into two distinct waves.[7] A disturbance viewed as noise in the isotropic model becomes a predicted observation in the anisotropic one. This is an example of the theory-ladenness of observation, introduced in Section 1.3.3, above, and discussed further in Section 2.1.4.3, below. This is also an example of applicability of different models within seismology.

Moreover, an interesting problem arises if we consider the statement that P waves are faster than S waves, from a Positivist point of view. For an observational seismologist, P and S denote the primary and secondary disturbances travelling through the Earth and recorded at the receiver. Thus, it would appear that the above statement is true by definition. However,

---

[5] *see also*: Slawinski (2016, p. 4)
[6] *see also*: Slawinski (2015, Sections 6.1.1–6.1.3)
[7] *see also*: Slawinski (2015, Chapters 7 and 9)

since P and S are derived from differential equations, we can prove that the P wave is at least $2/\sqrt{3}$ times faster than the S wave.[8] Hence, the fact that the P wave arrives ahead of the S wave is entailed by the equation of motion within an isotropic Hookean model.

2.1.2.3 *Verification*

[9]Let us turn now to the verification principle, which appears in different degrees of strictness. The most stringent approach demands that a direct observation be actually feasible. A less strict version would allow the statement that the outer core is liquid to be meaningful, if we could suggest a method, say, of drilling to its hypothesized depth. Even though performing such a drilling might not be practically possible, merely suggesting it would be, in principle, sufficient. The statement would be meaningful; it would be true or it would be false, though we may never know which. Positivists considered religious sentiments as instances of meaningless statements, since we do not know, even in principle, what might count as empirical evidence for the claim that God loves us.

2.1.2.4 *Predictions and explanations*

The role of theoretical entities in the Positivist picture is to organize and systematize observable experience. This is different from theorizing in the hope of understanding the process. Positivists were not keen on looking for the hidden causes of things. Indeed, in their view it is generally meaningless to talk about such things. Nevertheless, there was an elegant theory of explanation, due to Carl Hempel, that agreed with the rest of the Positivist outlook.

An explanation, according to Hempel (1970), is an argument for which the theory and the initial conditions are premises. The explanation is deduced as a conclusion. Consider an explanatory answer to the following question. Why is there an eclipse of the Sun at a given time? Suppose, we state the theory that describes the motion of the planets around the Sun and of the Moon around the Earth. We also need an auxiliary theory that light travels in straight lines and can be blocked by a material body. The initial conditions are the positions and velocities of the Earth, Moon, and Sun at an earlier time. From these premises, we derive the conclusion. There is an eclipse of the Sun at a given time.

---

[8] *see also*: Slawinski (2015, Exercise 5.18) and Slawinski (2016, Exercise 5.3)
[9] *see also*: Slawinski (2016, Section 9.4)

Explanations are seldom, if ever, presented in such a formal way, but—if so desired—they can be expressed in a canonical form. Seismologists could generate explanations such as the following.

1. Theory: The Earth has inner and outer cores of specified sizes.
2. Theory: Elastodynamic theory describes propagation of disturbances within the Earth.
3. Theory: An explosion within the Earth generates a disturbance.
4. Initial condition: There is an explosion at a given location and at a given time.
5. Auxiliary theory: There is a relation between the propagation speed of a disturbance and rock properties.
6. Initial condition: Rock properties are known.

o Conclusion: A sensor at a given location records a disturbance at a given time.

Herein, there is a symmetry between prediction and explanation. If we follow such an argument after the eclipse has been observed or the explosion detected, we are explaining it. If we perform the derivation prior to observation or detection, we are predicting the event. The elegance of this symmetry cannot be doubted. Also, since prediction is tied to confirmation—that is, making correct predictions is a symptom of the theory being promising— there is an intimate bond among explanation, prediction and confirmation. Such an account of explanation is intellectually attractive.

Elegant as the Hempel (1970) account is, it has many problems, as illustrated by the following example.

Example 2.1. Flagpole
Consider a flagpole of a certain height casting a shadow of a certain length. Let us explain this length. The location of the Sun, the nature of light and shadows, together with trigonometry and the knowledge of the height imply that length, which is the sought explanation. Now, let us try to explain the height of the pole. We begin with the known length of the shadow and then using the other information, we calculate the height of the flagpole. However, this is not an explanation, since a notion of causation is also required. Flagpoles cause shadows, but shadows do not cause flagpoles.

As stated by the title of Thom (1993), *Prédire n'est pas expliquer.*[10]

---

[10] *To predict is not to explain.*

An issue that is left out of Hempel (1970) is the concept of understanding. Explanations appear to be objective, but understanding is a psychological state. In quantum mechanics, it is logically possible to have an explanation, according to Hempel (1970), that is humanly unintelligible. Explanation is an aim of science, but so is understanding. They are related, but their relation is not obvious, and hence, it is debated (Woodward, 2009).

2.1.2.5 *Contentious issues*

Positivism's appeal was, at least in part, a result of its exceedingly optimistic, and perhaps even naive, perception of the subtleties of science. The verification principle was a powerful method to recognize nonsense. Empirical science was acknowledged while pseudoscience and dogmatic political doctrines were found to be nonsensical. Those were appealing qualities of Positivism. Yet, several branches of knowledge were in danger of being ignored. Ethics, mathematics, and philosophy itself were problematic. None of them is open to empirical testing, so each seemed to lack cognitive significance.

The Positivist attempted to find a solution to the problem of mathematics in language. Certain statements, such as 'John is a bachelor', report matters of fact. Others tell us nothing about the world, but instead report the use of words, such as that 'all bachelors are unmarried men'. We can verify that the second statement is true by looking in a dictionary, since its truth is wholly dependent on the meaning of the terms. To verify the first statement requires an empirical inquiry to see, for instance, if John and his partner have been through a particular legal process. Mathematics, according to a Positivist account, is based on the meaning of the terms involved. For instance,

$$5 + 7 = 12$$

is a true statement because of the meanings of numbers $5, 7, 12$, operation $+$ and symbol $=$, not because there is an independent realm of facts on which mathematicians rely.

Science, according to the Positivist outlook, organizes our experience and predicts our observations. It does not, however, attempt to gain an insight into causal explanations. While many who share Positivist instincts think that such an outlook is largely correct, realists strive to go beyond appearances and attempt to grasp reality. They strive for explanations and understanding, which extend beyond the organization and prediction of observations. Positivists are antirealists, which means that, according to them, science only aims at theories that are empirically adequate. In other words, theories account for observations, but do not allow us to infer further information about the reality hidden to an observer. Since the theoretical terms are only indirectly defined through observational terms, or possibly not defined at all, there is little point in claiming that theoretical terms grasp the physical essence of the world. Such an antirealism has a long and distinguished history.

In astronomy, it has been often maintained that since we have no access to the sky, it is hopeless to seek its intrinsic properties. We can hypothesize the celestial mechanics, but solely to predict our observations. According to such a view, we strive for a reliable instrument, not the truth.

2.1.3　*Popper and falsification*

2.1.3.1　*Popper's approach*

Karl Raimund Popper, introduced in Section 1.4.2, was the chief rival of the Positivists, especially during the middle years of the twentieth century. He took a very different view of theory evaluation. Theories often have the form of all *A*s are *B*s, such as that all ravens are black or that all bodies attract one another with a force proportional to the product of their masses and inversely proportional to the square of their distance apart. To refute such a theory one needs only a single instance of *A* that is not *B*. A black swan refutes the theory that all swans are white. However, to confirm such a theory, Popper asserts, one would have to check every single instance of *A* to ensure that it is also *B*. This approach is impractical and, in many instances, impossible. It would require an exhaustive search of the entire universe, past, present, and future. From a logical viewpoint, decisive refutations are possible, but confirmations are not. Those who seek confirmations rather than refutations lack the critical spirit, says Popper.

According to Popper, the proper method of science is conjectures and refutations. It is hopeless to try to confirm any theory, but we can make scientific progress through trial and error.

Furthermore, Popper was interested in defining the boundary between science and nonscience and the criteria for that boundary. He called it the demarcation problem and demarcation criterion, respectively. His solution appeared simple. To be scientific is to be refutable in principle. That is, the difference between a genuine scientific theory and any piece of pseudo-science lies in its openness to refutation. A theory must make claims about the world that, for all we know, could turn out to be false. A seismological theory that predicts an earthquake next Tuesday afternoon is open to refutation. If there is no earthquake, the theory is refuted.

Popper was an empiricist, but he approached the issue in a manner considerably different from the Positivists. According to him, concepts are taken to experience, not derived from experience, and there is no significant distinction between the observational and theoretical concepts. As a scientific entity, a collected fossil and the Earth's inner core are not different from one another. Science makes conjectures about both of them, and these conjectures are open to refutation. Popper was also a common-sense realist; he took scientific theories to be claims about how things really are, not just useful instruments for organizing and predicting experience.

The essence of Popper's approach is the following. Scientists are—or ought to be—intellectually honest and thus willing to take defeat when it comes, as it almost inevitably does. Popper's picture is contrary to the image of the cautious scientist who diligently collects data and then guardedly makes only the most circumspect inferences. According to him, science is revolutionary. It is not the cumulative acquisition of carefully established facts. Furthermore, it is more than the organization of observable experience. Science is an unending search for an explanation and understanding of reality. While many scientists embrace such a view, philosophers are more hesitant.

Many scientific activities do not fit Popper's approach. Certain activities are not attempts to explain, but are rather a search for interesting new phenomena. Also, Popper's approach ignores the fact—emphasized by Thomas Kuhn (1996), whose views are discussed in Section 2.1.4, below—that much of science is puzzle-solving: an attempt to make things fit into an already accepted pattern of explanation.

There is no strategy for, or logic of, discovery, according to Popper. The only place where rationality can play a role is in the testing of theories after they have been proposed. It is ironic that Popper's most famous book, *The Logic of Scientific Discovery* (Popper, 1959), denies the very existence of the subject matter of its title.

However, the distinction between discovery and justification is a point on which Popper and his Positivist rivals agree. They claim that background beliefs are irrelevant in the process of theory testing, since all bias can be filtered out in the process of justification and rigorous testing. However, it appears from history of science that a whole community can, knowingly or not, exhibit systematic biases. The belief that nature abhors a vacuum, *horror vacui*, was a bias accepted by a scientific community to explain a variety of observations (Close, 2009).

A bias held by an entire community is likely to be built into every theory and, hence, it might not be recognized through a process of justification. To avoid such a predicament, it might be advisable to formulate a wide range of diverse theories, since contrasts among them might render a bias visible.

Another problem is that, according to Popper, a single counterexample refutes a theory. However, theories do not, by themselves, imply observations. They do so in conjunction with further assumptions and auxiliary theories; the former might have a form of initial conditions and the latter might be necessary to calibrate the experimental apparatus, such as the electromagnetic induction used to record the motion of a seismograph. A lack of agreement between theory and observations does not specify the source of disagreement. In other words, it is impossible to isolate issues of the main theory, since its empirical tests require auxiliary assumptions and hypotheses.

A disagreement could stem from a fundamental conflict between observations and the main theory, but it also could be a result of false auxiliary theories or wrong initial conditions. This issue is called the Duhem problem or the Duhem-Quine problem, in honour of Pierre Maurice Marie Duhem, a French physicist, mathematician and philosopher of science, and Willard Van Orman Quine, an American philosopher and logician.

In seismology, examining a discrepancy between predictions and observations, we could—apart from the main theory—question the accuracy of measurements or the choice of an auxiliary theory. The latter might suggest the change of a model, say, from elastic to viscoelastic, or an adjustment of parameters within elasticity itself. Thus, Popper's refutability is not easy to apply.

In spite of pragmatic difficulties with refutability, Popper's most famous doctrine is falsifiability, as described at the beginning of this section. A theory that cannot, in principle, be refuted is not part of science. This rules out as nonscience such claims that there is life elsewhere in the universe. Since this claim does not say where or when this life exists, it would take an exhaustive search to refute the claim. This is not possible, so the claim is not refutable, and therefore it is not scientific, according to Popper.

There are insightful consequences of falsification, but as a methodological prescription it is too strong. Certain explanatory theories that unify their discipline, such as plate tectonics, might not make predictions that are specific enough to apply the criterion of falsifiability. According to Popper, any form of inductive confirmation is hopeless, and positive evidence in support of a theory is an illusion. This seems excessive. Making sense of induction is a central part of understanding the crux of any scientific method. According to Popper, the scientific spirit is an unsparing critical spirit. This—to a certain extent—might be true. Yet, commonly, scientists attempt to make adjustments within the accepted paradigm, rather than to overthrow the theory, as discussed in Section 2.1.4, below.

2.1.3.2 *Popperian seismology*

Working scientists often find Popper's view attractive and use it in scientific debates. String theory, for instance, has champions and opponents based on Popper's approach to science with emphasis on testable consequences. Lee Smolin (2007) explicitly appeals to Popper in his condemnation of string theory. We must, he insists, have testable theories; anything else gets in the way of scientific progress.

Albert Tarantola (2006) suggests that seismologists could consider a falsification in a form modified by probability: a probabilistic falsification. According to Tarantola, a model or a theory is refuted if it does not meet a required threshold of likelihood. This is a natural extension in the spirit of Popper.

This point is important when considering different types of theories. Quantum mechanics is intrinsically a probabilistic theory; for instance, it predicts a range of decay times for a radioactive material. No single instance could falsify this theory; it would take a great many that fell outside the predicted range. In this case, Popper's falsification could not be applied except in a probabilistic form.

Seismology is intrinsically a deterministic theory; it strives to make specific predictions that, in principle, fit the standard Popper criterion. However, probabilistic considerations come into play on two accounts. First, they appear if we take into account measurement errors. Our measurements might result in distributions, which entail probabilistic considerations, but in a manner different from quantum mechanics. In the quantum case, the results are—we might say—due to the laws of nature, while in the seismological case, the distribution is due to inaccurate measurements. Second, probabilistic considerations appear due to the fact that inferences in seismology are mediated by continuum-mechanics models, and different models exhibit different agreements with measurements. If we consider a probabilistic distribution of a value of a given parameter of a Hookean solid, we consider both the measurement errors and the pertinence of that solid as model.

2.1.4 *Kuhn and paradigms*

2.1.4.1 *Kuhn versus Popper*

The term 'paradigm' and the phrase 'paradigm shift' entered popular culture through the nomenclature of Thomas Kuhn, one of the most influential intellectuals of the twentieth century. His remarkable book, *The Structure of Scientific Revolutions* (Kuhn, 1996), attempts to show how science is done. Philosophers are largely concerned with how science ought to be done, while Kuhn (1996) aims to describe actual scientific practice. However, the two formulations should not be exceedingly different from one another. If scientists are irrational according to criteria of a normative philosophy of science, which is a proposed account of scientific method, then this should be seen as a *reductio ad absurdum* of that philosophical theory. This is why Kuhn's book is so important to both historians and philosophers of science.

Kuhn (1996) has a couple of issues to address. Against the Positivists, he maintains that the history of science has not been solely cumulative; there have been a number of significant revolutions that overturned the most basic beliefs of earlier scientists. Against Popper, he argues that science has not been characterized by incessant criticism, but—for extended periods—it proceeds in an inductive and cumulative manner that leaves foundational issues unquestioned.

An example of purported cumulativeness is Newton's theory of gravitation, which generalized and explained Kepler's laws of planetary motion. Thus, Newton's laws of gravitation and motion contain Kepler's laws. In seismology, the aforementioned P and S waves are derived from the elastodynamic equation constrained by the assumption of isotropy, which is directional independence of properties of a medium through which waves propagate.[11] Upon relaxing that assumption, we find that there are three types of waves in anisotropic media, as mentioned on page 26.[12] As in the case of Newton and Kepler, the anisotropic formulation contains the

---

[11] *see also* : Slawinski (2015, Sections 6.1.1–6.1.3)
[12] *see also* : Slawinski (2015, Chapters 7 and 9)

isotropic one. According to that view of science, subsequent theories are mere generalizations that contain their predecessors.

By contrast, Popper held that the history of science is a history of conjectures that are submitted to tests. Any theory that does not pass a test is rejected. The history of science is not cumulative but revolutionary; it is littered with refuted theories.

2.1.4.2 *Kuhn's account of science*

Kuhn rejects both the Positivist and Popperian pictures of science. He accepted that science is not cumulative, but revolutionary. However, it is not constantly revolutionary; there are periods during which criticism stops. Kuhn's account of scientific change includes a number of important details.

Normal science   As the title suggests, most science is normal science. Scientists agree on the basics, which are taken for granted; the fundamental theoretical framework is beyond criticism. Normal science is a puzzle-solving activity. It is an attempt to fit things into a pattern that is set by a paradigm.

Paradigms   Kuhn's most famous contribution was that of a paradigm. There are several concepts associated with a paradigm. Among them, we have ontology, which describes the world, heuristic norms, which guide us through the process of normal science, and evaluative norms, which are the criteria for judging proposed solutions. Within a given paradigm the science is cumulative.

Crisis   Certain problems among the puzzles and anomalies encountered within a given paradigm might appear insoluble. If many of them persist for a long time, a period of crisis arrives. In agreement with the proverb that it is a poor worker who blames his own tools, the admission of crisis comes with reluctance. Crisis induces attacks on the paradigm itself; basic assumptions are questioned. Scientists begin to consider anomalies not as puzzles but as counterexamples to the theory.

Scientific revolution   Normal science is no more; the community is disintegrated into several diverse schools. Several prototypes of paradigms come into existence in this period of extraordinary science. Eventually one of them prevails, and the majority of scientists accepts the new theory. This is a scientific revolution.

Entrenchment of revolution   The way a revolution in science becomes entrenched is not by further empirical evidence or argument, but rather by indoctrination. Students learn the new theory from textbooks that present the exemplars that illustrate the theory as if they were independent evidence for it. Textbooks often give a short history of the subject that makes the accepted paradigm look reasonable and its rivals look silly. Consider how Galileo's rivals, caricatured in Figure 2.1, are depicted as dogmatists who refused to look through his telescope.[13]   Much as it is for national histories, it is the winners who write the science books.

Fig. 2.1

Kuhn is not critical of the attitude that leads to the entrenchment of ideas. He considers it normal for practicing scientists not to be skeptical about basics.

Young scientists learn by example in the laboratory and much of what they learn is a skill of knowing how, as opposed to knowing that. This non-propositional knowledge is important and plays a role in doing science. One

---

[13]Readers interested in scientific arguments of Galileo's rivals might refer to Graney (2015).

can ride a bicycle without understanding its physical principles. Be that as it may, a new generation grows up familiar with a new paradigm, and we are again in a normal science, often without understanding its foundations.

2.1.4.3 *Contentious issues*

There are several contentious issues that arise in Kuhn's account of science. Thus, philosophers and historians of science have taken Kuhn to task on many fronts though not before conceding the extent of Kuhn's extraordinary contribution.

Fig. 2.2

Theory-ladenness of observation   According to Kuhn, perception is paradigm-dependent. Our observations are never neutral. Seeing is always 'seeing as' or 'seeing that'; it is never a passive observation, since our beliefs and expectations condition what we see. Nothing is independent of a conceptualization, and—consequently—there can be no independent empirical basis for testing rival theories. As illustrated in Figure 2.2, if we expect to see a duck, we do but if we expect to see a rabbit, we see that instead. Herein, we readily distinguish between two paradigms, but it might be difficult in more sophisticated structures.

Meaning and incommensurability   According to Kuhn and the Positivists, the meaning of a term is dependent on the role it plays in a given theory. Even though Kuhn rejected the Positivist's distinction between theory and observation, he accepted the view of language in which key terms are defined contextually. Their meaning is obtained from the role they play within the theory. Consequently, a change of theory involves a change in meaning. Consider the term mass, which might have different meanings in classical mechanics and in relativity. Within distinct paradigms, scientists may appear to say contradictory things about mass, but actually they are invoking different concepts. The different theories are incommensurable with each other, and there is no common language in which to express them.

Example 2.2. Duck-rabbit paradigm

Consider the distance from eye to mouth in the rabbit and duck paradigms illustrated in Figure 2.2. The distances do not agree with one another, because the meaning of mouth and, hence, the distance in question are paradigm-dependent.

History   Kuhn's claims that normal science is dominated by a single paradigm, such as the present dominance of plate tectonics in the geosciences. However, the history of science shows periods of coexisting rivals.

Revolutionary tools of science   Also, important features of science that are missing from Kuhn's account are the revolutionary tools and methods that are not new theories. Computers, for instance, have had a revolutionary impact on science. Dyson (2001) emphasizes this point.

> The concept-driven revolutions are the ones that attract the most attention and have the greatest impact on the public awareness of science, but in fact they are comparatively rare. In the last five hundred years we have had six major concept-driven revolutions, associated with the names of Copernicus, Newton, Darwin, Einstein and Freud, besides the quantum-mechanical revolution that Kuhn took as his model. During the same period there have been about twenty tool-driven revolutions, not so impressive to the general public but of equal importance to the progress of science. I will not attempt to make a complete list of tool-driven revolutions. [...] The effect of a concept-driven revolution is to explain old things in new ways. The effect of a tool-driven revolution is to discover new things that have to be explained. In physics there has been a preponderance of tool-driven revolutions. We have been more successful in discovering new things than in explaining old ones.

In seismology, the revolutionary effect of numerical and computational techniques has allowed us to extract subtle information contained within data, as developed—in the context of exploration seismology—by Robinson and Treitel (2000). Also, and perhaps more importantly for theorists, equations can be solved numerically, without simplifying assumptions required for analytical solutions.

For instance, accounting for seismic information without assuming isotropy of a medium is, at least in part, a consequence of the computer revolution. It is also a consequence of an increased accuracy of measurements, which is a technological revolution.

Furthermore, mathematical methods created for their own sake became practical tools thanks to the availability of computers. An example of such a method is the Radon transform, formulated in 1917 by Johann Radon, which is an essential tool for image creation in both seismological and medical tomography. Application of this transform requires computational resources not available at the time of its formulation.

Seismology benefits from both concept-driven and tool-driven advances. Commonly, the concept-driven and tool-driven advances stimulate one another, as exemplified by the discovery of the inner core, which is discussed in Example 1.4, and by the formulation of seismic ray theory.

2.1.5 *Historical approaches*

The motivation for historical approaches is consistent with the Kuhnian approach, where history can teach us a lot. According to Kuhn, Popper and the Positivists are too *a priori* in their approach to the norms of science and do little justice to actual history. In abandoning those approaches, we still wish to perceive science as a rational activity. The leading figures in this alternative approach include Imre Lakatos and Larry Laudan.

The methodology of scientific research programmes (Lakatos, 1970) posited a basic entity that is similar to Kuhn's paradigms. Therein, a research programme is more general than any particular theory; it is a series of related theories. Such a programme has a so-called hard core, which is a set of fundamental principles that define it. It also has a protective belt, consisting of auxiliary assumptions that can be modified and take the blame if something goes wrong. There are no crucial experiments to test a theory, since we can always blame a false prediction on something in the protective belt rather than the fundamental principles.

The discovery of the planet Neptune is an insightful illustration. The hard core of Newtonian astronomy contains Newton's three laws of motion and the law of universal gravitation. These principles—and nothing else—could explain all planetary phenomena. At the time, it was observed that Uranus was not following the predicted orbit. A strict Popperian would take this as a refutation of Newton's theory. Lakatos claimed that the hard core is protected by additional assumptions that can be altered. In this case, the assumption that there are only seven planets was abandoned and an additional planet—so far unseen and eventually called Neptune—was hypothesized to be gravitationally interacting with Uranus. This led to predictions of where to look to find this new planet, which were eventually confirmed: a great triumph for the Newtonian research programme.

The idea of a protective belt, which prevents outright falsification, is a point that separated Lakatos from Popper. Lakatos, at first an avowed Popperian, initially claimed to be proposing a version of Popper that could accommodate aspects of Kuhn's criticism. But Popper claimed that Lakatos's views were fundamentally at odds with his own. There was, sadly, a professional and personal break between them before Lakatos's premature death in 1974.

Seismology offers excellent examples of Lakatos's account. In particular, the notion of a protective belt is prominent due to a variety of issues that can be invoked to justify the failure of prediction or explanation. Among issues protecting the accepted principles of seismology, we might consider the limited accuracy of measurements, the scarcity of experimental data, and theoretical approximations within continuum mechanics. Furthermore, twenty-one parameters of a Hookean model give us an ample opportunity to accommodate a large variety of experimental measurements even though the concept of a Hookean solid might not be appropriate as a model. In other words, such a large number of parameters makes the theory flexible and accommodating, but—on the other hand—it renders it difficult to overturn.

According to Lakatos (1970), research programmes that make the occasional correct novel prediction are progressive, while those that are constantly modifying themselves to keep up with new information are degenerating. Rationality in science consists in adopting progressive programmes and abandoning degenerating ones. As illustrated by the success in the discovery of Neptune, scientists should not jump ship too soon. Not surprisingly, Lakatos was unable to say how long a research programme had to degenerate before it became irrational to follow it. Without some guidance on this point, it is impossible to say when a scientist begins to act irrationally. This is a major stumbling block for this view.

To gain further insight into Lakatos's account in the context of seismology, consider three levels of seismological assumptions. All are fallible, but their status varies considerably.

First—at the level of a basic theory—seismologists make a choice of studying phenomena within the realm of continuum mechanics, as opposed to, say, condensed-matter physics. Saying that the principles of continuum mechanics are fallible means that this theory could be an inappropriate framework for a given study, but it need not be refuted.

Second—at the level of general principles, within the framework of continuum mechanics—seismologists might make a choice of the theory of elasticity, and use its principles without invoking, say, entropy. Fallibility at this level consists in mistaken assumptions about what can and cannot be ignored to maintain an adequate representation of phenomena.

Third—within the theory of elasticity—seismologists might make a choice of parameters that describe a given Hookean solid. Fallibility at this level consists in inappropriate values of parameters, which can be adjusted.

2.1.6 *Personal and social values*

There has long been a value-free ideal in doing science. Most of twentieth century philosophy of science espoused this idea. The claim of value-neutrality is that science is concerned with the facts and when research is properly done, values play no role in the resulting theory. It is often understood that if values are involved they have a corrupting influence.

Before proceeding, we need to distinguish two types of value. Cognitive values are an essential part of science, since they are the core of the scientific method. They are values because they are norms, not facts. Champions of value-free science do not intend to eliminate these types of values from science. Noncognitive values, however, such as social, political, cultural and religious, are the usual candidates for exclusion.

More sophisticated discussions admit that values—in the latter sense—play a role in the scientific process, but not in the content of a given theory. There are three stages of this process at which the value-based decisions appear.

First, there is a choice of research topic. This might involve a decision based on personal interests and availability of funds. Also, in this first choice, a person assumes, at least implicitly, that a given issue lends itself to a rational enquiry. The second stage involves the actual research work, whose result might be an improved theory evaluated by value-free evidence. The third stage involves the application of the results of research. Seismologists might infer objectively that there is oil at a given location, but the decision to extract it is a value-based choice. Thus, values play a role in what we choose to study and in how we use the results, but a scientific theory has no values in it.

The value-free ideal has inspired philosophers, scientists and the public. It is not tenable, however, for two reasons. One reason is that scientific method is filled with epistemic cognitive values. As mentioned above, there is a preference for simple theories over complex ones or for theories that make novel predictions. These values are essential to science; we could not proceed without their guidance.

There are also noncognitive values, and geosciences abound with examples. The debates between the catastrophists and the uniformitarians turned partly on empirical findings and partly on an assortment of values. Georges Cuvier thought that the transitions in flora and fauna documented by fossils were best explained in terms of catastrophic events. He was not particularly religious and made no mention of supernatural forces, but certain followers accepted the framework and tied it to Noah's flood. Empirical data was interpreted in terms compatible with a particular understanding of the Bible. In short, the theory was formulated and the data interpreted under the influence of religious values.

James Hutton and Charles Lyell rejected this in favour of uniformitarianism, which claimed that the Earth's features were the product of gradual processes taking place over millions of years. This view rejected the creation of Genesis, or at least reinterpreted it in nonliteral terms.

Regardless of our attitude, we must admit that the history of geosciences has been shaped by values. In relation to geophysics, Anduaga (2016, p. vii) states that

> science is inseparable from the society in which it is born, a
> society that—more than ever—depends on scientific advances.

Let us look at seismology as an example of the role of cognitive values in the content of scientific theories. We can describe the Earth as a Hookean solid. To do so, we choose a fourth-rank tensor to relate linearly the stress and strain tensors, which are—by their definitions—second-rank tensors. Such a formulation is not imposed *a priori* by either mathematical or physical considerations. For instance, there is no empirical evidence or mathematical structure that demands explicitly the choice of such a rank. If we chose an eighth-rank tensor, we would have a potentially more accurate theory, but one whose logistic complications might outweigh scientific benefits. With eighth-rank tensors, we would have more symmetries with which to consider various phenomena, but our predictions or retrodictions might be too subtle for any empirical support. Eighth-rank tensors would entail more detail in our hypothetical description of the Earth, but without justification by potential observables, especially if we consider errors in our data.

Theoretical choices affect our observations. If we use fourth-rank tensors, we interpret observed disturbances in an anisotropic solid in terms of three types of waves.[14] Anything else is considered as noise. If we use higher-rank tensors, then there are other types of waves entailed by a theory. Thus, a human choice of the underlying theory determines the types and number of waves interpreted from seismic data.

An instrumentalist, whose views are discussed below in Section 2.2.2, would say that the difference originates in models, and would accept the lower-rank tensors as empirically adequate. A realist, whose views are discussed below in Section 2.2.1, would agree that the difference originates in models, but would suggest that certain models might be closer to reality than others, even if we cannot make observations accurate enough to distinguish. For pragmatic reasons, a realist would agree to use lower-rank tensors if they offer the required adequacy. Herein, even if in a subtle manner, values play a role.

Finally, though we do not discuss it here, value can be objective. There is a difference between fact and value, but this is not equivalent to, respectively, objectivity and subjectivity. Under a set of reasonable criteria, we objectively value Bizet's Carmen as a great opera and Schubert's C-major cello quintet as a great chamber-music composition.

2.2 Ontological issues

2.2.1 *Realism*

2.2.1.1 *Realism and physical world*

Realism—or, more accurately, scientific realism—is a view that science aims to give us a true account of nature, which exists independently from us, and that science is largely, if not completely, successful in doing this. Most people would not give the matter another thought except for certain significant problems that put realism in doubt.

---

[14]*see also*: Slawinski (2015, Section 7.3)

For instance, the history of science is a history of overthrowing one theory after another. The pattern suggests that our current theories are endowed with a similar fate, which in turn casts serious doubt on the realist view that these theories are true, or at least approximately so. This argument is known as the pessimistic induction. Its conclusion is that science should not aim for truth, which is hopeless, but instead aim for empirical adequacy, that is, for predicting observations.

Another widely accepted reason to reject realism is to say that nature is ultimately inaccessible. This was a favourite argument of astronomers up to the time of Galileo. The underlying reality cannot be discovered, so the truth about that underlying reality should not be the aim of science.

There is no consensus among philosophers on the correctness or incorrectness of scientific realism, but there is a rough consensus on its characteristics. The aim of science is truth. The feasibility of this statement is due to two properties. First, the ontological aspect: the truth of a scientific theory depends on the physical world, and is independent of scientists. Second, the epistemological aspect: there is sufficient evidence—or at least the possibility of such evidence—for a theory, so that acceptance or rejection of it is based on that evidence.

Some opponents of realism reject the dichotomy between ontology and epistemology by linking the meaning of truth with actual evidence for truth. They find the idea of unknowable truth, which is a distinct possibility for a realist, to be absurd. Stebbing (1958/1937) questions such a rejection of dichotomy.

> Thus Nature, it seems, consists of our *perceivings* and of *that which we perceive*, indissolubly bound together. What, then, do we perceive?

She describes it as a confusion, where a scientist

> is unable to make up his mind whether *Nature* is our knowledge or whether our *knowledge* is *of* Nature.

2.2.1.2   *Realism in seismology*

Following Example 1.2, to argue for a realist interpretation of seismology, let us consider and explain the statement that there is a liquid outer core approximately 2900 kilometers below the surface (Brush, 1980). In 1926, Jeffreys used the Chandler wobble and tides of the Earth to infer the existence of a liquid core. About a century earlier, Siméon Denis Poisson showed the existence of the P and S waves in a Hookean solid. In 1906,

Oldham, detected two types of disturbances from earthquakes, which he interpreted in terms of P and S waves. In doing so, he idealized the Earth as a Hookean solid. Presently, the pattern of disturbances observed at various points on the Earth's surface is consistent with the existence of a liquid core.

Even though both approaches share common assumptions, such as the spherical symmetry of the Earth, and both use continuum mechanics as the tool of investigation, their common conclusion—derived through different approaches—strengthens the statement about the existence of the outer core.

To gain an insight into such strengthening, consider the following argument. If an approach is false, then it is unlikely that it makes accurate predictions. If two approaches make the same prediction, or retrodiction, their falsity and, hence, the falsity of their conclusion is even less likely.

The discovery of the liquid core provides us with an even stronger argument for realism, if we consider it in stages. Jeffreys's conclusion made possible the refinement of the seismological interpretation of the data. Without Jeffreys's conclusion, the argument for the liquid core and its dimensions—based on sparse seismic data alone—was too weak. The assumption of the liquid core—understood realistically—is essential for this result.

Furthermore, the success of the approach initiated by Poisson and followed by Oldham suggests that a Hookean solid is a fruitful model for investigations of the Earth. Different elements fitting together support the realistic outlook.

2.2.2 *Instrumentalism*

Antirealist views are many and varied. Herein, we describe instrumentalism, with its relevance to the conceptual issues in seismological theory.

One of the purposes of instrumentalism is to avoid the problems of underdetermination brought to our attention, in the context of physics, by Duhem (1914, 1991), and referred to as the Duhem problem, discussed on page 32. John Stuart Mill (1919) acknowledges the importance of this problem by stating that

> most thinkers of any degree of sobriety allow that an hypothesis of this kind is not to be received as probably true because it accounts for all the known phenomena, since this is a condition sometimes fulfilled tolerably well by two conflicting hypotheses; while there are probably many others which are equally

possible, but which, for want of anything analogous in our experience, our minds are unfitted to conceive.

According to instrumentalism, the aim of a scientific theory is empirical adequacy, not truth. Theories can organize and predict observations. However, a description of an underlying reality is useless, irrelevant and—in principle—unattainable.

Strictly speaking, a seismological theory might invoke terms such as the liquid outer core, but these are taken only as a vehicle to account for sensory experiences, such as geophysical data. The only statements that are taken to be true are those with observational terms, such as disturbances within seismic detectors at various locations on the Earth's surface.

Parts of seismological studies have aspects of instrumentalism, *sensu lato*. For instance, to study a particular region within the Earth with an attitude of a realist, seismologists might treat the surrounding areas instrumentally, only to accommodate the behaviour of seismic waves within them. Thus, as illustrated in Figure 1.4, to study the core, seismologists must consider waves that travel through the mantle. However, if the purpose of study is the core itself, it is common to treat properties of the mantle in a manner that allows us to accommodate the time it takes the wave to propagate through it, without the need of a realistic examination of its mechanical properties.

Also, most of applied seismology, whose purpose is the exploration and exploitation of natural resources or an evaluation of stability of underground material, could be viewed as an instrumental science, *sensu lato*. A seismologist engaged in such studies might use seismic data with the same philosophical attitude as a mariner would use patterns of stars for navigation. In both cases, the purpose of science—and, hence, criteria for its success—are outside of it. The choice of a geocentric versus heliocentric formulation of astronomy or an isotropic versus anisotropic formulation of ray theory is only pragmatic; such a choice has no other meaning, provided that the mariner's destination is reached, the anticipated resource is found or the required stability is achieved.

The problem of underdetermination of theory by evidence is distinct from the issues of relating physical concepts to analogies provided by mathematical structures, where different structures can be used to examine the same physical phenomena, as exemplified by the Newtonian, Lagrangian and Hamiltonian formulations of classical mechanics (Arnold, 1989; Goldstein, 1980), and the same structure can be used to examine different phenomena, as exemplified by equation (3.4) on page 72, below. These issues belong to Platonism, and do not imply an antirealist view, even though, as stated by Henri Poincaré (1968, Chapter XII),

> [l]es théorie mathématiques n'ont pas pour objet de nous révéler la véritable nature des choses; ce serait là une prétention déraisonnable. Leur but unique est de coordonner les lois physiques que l'expérience nous fait connaître, mais que sans le secours des mathématiques nous ne pourrions même énoncer.[15]

Herein, Poincaré (1968) opposes the Platonic approach to, but not a realistic view of, physics.

Fig. 2.3

Closing remarks

Foundational issues concerning science are likely to remain subjects of debate. For instance, Voltaire (1734)—referring to the voyage of de Maupertuis and Clairaut to Lapland to make measurements for examination of the

---

[15]The purpose of mathematical theories is not to provide the true nature of things; it would be an unreasonable pretense. Their sole purpose is to organize the physical laws that experiments allow us to obtain, but which we could not even formulate without the help of mathematics.

polar flattening of the Earth, as illustrated in Figure 2.3—muses about the necessity of an empirical support for a theoretical conclusion.

> Héros de la Physique, argonautes nouveaux,
> Qui franchissez les monts, qui traversez les eaux,
> Ramenez des climats soumis aux trois couronnes,
> Vos perches, vos secteurs et surtout deux Lapones,
> Vous avez confirmé dans ces lieux pleins d'ennui,
> Ce que Newton connut sans sortir de chez lui
> Vous avez arpenté quelque faible partie
> Des flancs toujours glacés de la Terre aplatie.[16]

Beyond the foundational debates on epistemological and ontological issues, most branches of science have on-going debates about methodological issues within their fields. For the most part, these are not debates over the facts, but about the methods.

We can draw morals concerning methodology that applies to all the sciences, but our focus in subsequent chapters is on issues of primary concern to seismology. Terms such as realism, model, idealization and applied mathematics are commonly used in seismology as if they were clearly understood; however, each is problematic in its own way.[17]

---

[16] Heroes of Physics, modern Argonauts/Who scale the mountains, who traverse the seas/Who bring back from the climates under three crown/Your rods, your sectors and above all two Lapland women/You have confirmed in these nuisance-prone locations/What Newton knew without leaving his abode/You have surveyed some minor parts/Of permanently frozen flanks of the flattened Earth.

[17] Readers interested in the logical status of theories including realism, operationalism, hypotheticodeductive methods and falsifiability might refer to Hesse (2005/1961, Chapter 1).

Chapter 3

On continua and models

[...] conjectures [...] that are generally assumed by seismologists to be true, are properties of infinitesimal motion in classical continuum mechanics for an elastic medium with a linear stress-strain relation.[1]

Keiiti Aki and Paul G. Richards (2002)

Preliminary remarks

The theory that allows seismologists to describe the material, in which seismic phenomena occur, is the theory of continuum mechanics. Historically, conflicts between atomic theories and the concept of continuum have been a part of science since the ancient Greeks. However, the continuum is not postulated by seismologists as an attempt to answer fundamental questions regarding the structure of matter, but as a means to study mechanical properties of granular materials. In certain respects, it is akin to using averages instead of studying individual cases.

In contrast to the atomic description of matter, a mechanical continuum possesses—at any point—the same properties at every scale. These properties are assumed to be the ones of our everyday macroscopic scale, such as rigidity, compressibility, viscosity. Apart from allowing our everyday experience to bear on the formulation of the theory, such an assumption

---

[1]Similarly, Kennett and Bunge (2008, p. 1) write

> The development of quantitative methods for the study of the Earth rests firmly on the application of physical techniques to the properties of materials without recourse to the details of atomic level structure. This has formed the basis of seismological methods for investigating the internal structure of the Earth [...]

lends itself to the application of calculus, where continuity of functions is of paramount importance.[2]

In spite of its immediacy to our everyday experience, such a continuum is an abstract concept established to study the behaviour of real materials, in particular, their deformations. A mechanical continuum is an abstract entity that should not be confused with a physical material; nor should either be confused with the mathematical continuum.

We begin this chapter by examining the concept of continuum. We introduce the mechanical continuum in the context of the mathematical one. We discuss continuum mechanics, its primitive concepts and first principles. We conclude this chapter with an examination of the role of applied mathematics and related models and idealizations.

3.1   On mathematical continuum

To introduce the mathematical continuum, we can begin with the set of natural numbers, $\mathbb{N}$, namely, $\{1, 2, 3, \ldots\}$, and then, step by step define the real numbers, which form the analytic continuum.

First, we need the integers, $\mathbb{Z}$, namely, $\{0, 1, -1, 2, -2, 3, -3, \ldots\}$. The set of rational numbers, $\mathbb{Q}$, consists of the fractions. They are defined as equivalence classes of pairs of integers. We might try to represent rational numbers such as $1/2$ or $137/49$ as ordered pairs $(1, 2)$ and $(137, 49)$, respectively. The problem is that

$$\frac{1}{2} = \frac{23}{46} = \frac{512}{1024}.$$

They are the same fraction expressed with different terms. Thus, we need to construct rational numbers out of equivalence classes of pairs.

We might wonder if we need more than the rational numbers in physics. Every measurement is a rational number, which has a finite decimal expansion, plus an interval representing a measurement error. If r_1 and r_2 are rational numbers, then $(r_1 + r_2)/2$ is midway between them. This means that no matter how close two rational numbers are to one another, there is a rational number between them. In spite of this, there are still gaps in the number line, if all we have are the fractions.

For instance, according to the Pythagorean theorem, the length of the diagonal of a unit square is $\sqrt{2}$, and $\sqrt{2}$ is not equal to any number of the

---

[2]Readers interested in an insightful introduction to concepts of continuum mechanics might refer to Roberts (1994).

form p/q where p and q are integers. In other words, the point on the line that corresponds to $\sqrt{2}$ does not correspond to a rational number. The rational numbers are not complete.

Fig. 3.1

However, the real numbers, including $\sqrt{2}$, can be constructed out of the rationals by means of the Dedekind cut, which is playfully illustrated in Figure 3.1. Let us discuss the Dedekind cut using $\sqrt{2}$. We begin with a sequence of rational numbers,

$$S = \left\{ 1, \frac{14}{10}, \frac{141}{100}, \frac{1414}{1000}, \frac{14142}{10000}, \dots \right\} \tag{3.1}$$

that approaches $\sqrt{2}$. Numbers such as $15/10$ are greater than any member of S. They are its upper bounds. The least upper bound of a set of numbers is the smallest real number such that no member of the set is greater than that number, which might itself be in or out of the given set. For instance, S does not contain a least upper bound. Dedekind's axiom of completeness states that any set of real numbers with an upper bound has a least upper bound. The least upper bound of S is precisely $\sqrt{2}$.

Richard Dedekind used this idea to construct the real numbers from sets of rational numbers. Namely, he defined the aforementioned cut to be any partition of all rational numbers into two nonempty sets, A_- and A_+, such that every member of A_- is less than every member of A_+. Some cuts correspond to rational numbers. For example, take for A_+ the set of all rational numbers greater than or equal to $1/2$, and for A_- the set of all rational numbers less than $1/2$. This cut corresponds to $1/2$, which is a rational number. But there are other cuts.

For example, take for A_+ the set of all positive rational numbers whose square is greater than 2; in other words, $x > 0$ and $x^2 > 2$. For A_- take the set of all rational numbers that satisfy $x < 0$ or $x^2 < 2$. Thus, all negative rational numbers as well as set (3.1) are contained in A_-. Note that both A_- and A_+ are formulated solely in terms of rational numbers and their operations. Since there is no rational number whose square is equal to 2, every rational number belongs to either A_- or A_+, as required for a partition. This cut does not correspond to a rational number, so we identify it with a new entity, denoted by $\sqrt{2}$. In general, the cuts allow a construction of the real numbers, $\mathbb{R}$, from rational numbers, $\mathbb{Q}$, and their operations.

All real numbers can be identified with the Dedekind cuts. One can define what it means for one cut to be less than another and also the operations of addition and multiplication on cuts. It can then be formally shown that the $\sqrt{2}$ cut, defined above, is greater than 0 and satisfies $\sqrt{2} \times \sqrt{2} = 2$. The techniques that Dedekind introduced allow us to recover all the familiar properties of real numbers and place them on a firm foundation. This is the analytic continuum.

Apart from its foundational importance, the Dedekind cut appears, at least implicitly, in real analysis. For instance, a common proof of the intermediate-value theorem relies on the Dedekind cut.

Dedekind's theory of real numbers is one of the great achievements of nineteenth century mathematics. Another one is Georg Cantor's discovery of the transfinite hierarchy.

In common sense, infinity is thought to be something that is unending. The set of natural numbers is an infinite set, because we can count for ever, never coming to the last number. Prior to Cantor, the infinite was thought to be paradoxical. For instance, there should be only half as many even numbers as there are integers, yet both sets are infinite. Dedekind proposed the following definition.

Definition 3.1. A set, S, is infinite if and only if it can be put into one-to-one correspondence with one of its proper subsets.

Thus, the set of natural numbers is infinite since it can be put in one-to-one correspondence with the even natural numbers. An equivalent definition is the following.

Definition 3.2. S is infinite if and only if it has a subset that can be put into one-to-one correspondence with natural numbers, $\mathbb{N}$.

We can then ask about the size of other sets, such as the rational numbers, $\mathbb{Q}$.

Definition 3.3. Two sets are said to be of the same size if they can be put into one-to-one correspondence with one another.

Therefore, the set of all integers is of the same size as the set of just even integers. Also, it turns out that $\mathbb{Q}$ is of the same size as $\mathbb{N}$. Furthermore, any interval is of the same size as $\mathbb{R}$.

Are all infinite sets the same size? Is there only one infinity?

Cantor showed that the answer is no, by introducing the technique of diagonalization. Is the continuum, $\mathbb{R}$, the same size of infinite as the natural numbers? The answer would be affirmative, if we could put the two sets into one-to-one correspondence. If the answer is no, then—provided the two sets are comparable, which requires the axiom of choice—there are two possibilities. $\mathbb{R}$ could be smaller than $\mathbb{N}$, but this cannot be true, since the natural numbers are part of $\mathbb{R}$. Thus, the second possibility must be true: $\mathbb{R}$ is larger than $\mathbb{N}$.

Theorem 3.1. $\mathbb{R}$ *is larger than* $\mathbb{N}$.

Proof. Let us assume that $\mathbb{R}$ and $\mathbb{N}$ are of the same size, and hence, can be put into one-to-one correspondence. In other words, all real numbers can be arranged into a list, $r_1, r_2, \ldots$. The ordering of the list is irrelevant. We only assume that every one of them is on the list. It follows that the real numbers between 0 and 1 can also be listed in some way, for example,

$$1 \longleftrightarrow 0.\mathbf{6}5347\ldots$$
$$2 \longleftrightarrow 0.8\mathbf{2}471\ldots$$
$$3 \longleftrightarrow 0.33\mathbf{8}42\ldots$$
$$4 \longleftrightarrow 0.917\mathbf{2}6\ldots$$

$$\vdots$$

Let us define a real number using the diagonal digits in the decimal expansions of the real numbers on our list. We take the first digit in the decimal expansion of the first number and change it, say, according to the following rule. If the digit is not 2, then we replace it by 2; if the digit is 2, then we replace it by 1. Then, we change the second digit in the decimal expansion of the second number according to the same rule, and so on. This gives us a new real number $r = 0.2121\ldots$. Though this is a real number, it cannot be found on the list since it differs from each number in at least one decimal place. Thus, the list is incomplete. Indeed, it is essentially incomplete, since adding r to the list would not help. We could make another diagonal argument showing another number that is also not on the list. Since $\mathbb{R}$ is neither smaller than nor the same size as $\mathbb{N}$, it follows that $\mathbb{R}$ is a larger infinity than $\mathbb{N}$.   □

The analytic continuum serves as a model for other continua. The one-dimensional geometric continuum has, at least locally, the same structure as the analytic continuum, $\mathbb{R}$, and the n-dimensional continuum has the structure of $\mathbb{R}^n$. The same can be said of the mechanical continuum; it, too, has the structure of the real numbers.

Thus, we have the set of real numbers, which is the analytic continuum, geometric entities, such as lines and planes, which constitute geometric continua, a smooth structure of the mechanical continua, such as Hookean solids, which serve as models for physical continua, such as the Earth illustrated in Figure 1.4. None of them is part of the physical realm, but they are used to study that realm.

Herein, the physical continuum invokes the mechanical continuum to represent phenomena in the context of seismology. A specific example of the physical continuum is the Backus (1962) average, which represents a mechanical continuum as it appears to a seismologist who examines it

Fig. 3.2

within the constraints of the long-wavelength resolution available with seismic data.[3]

In view of several distinct continua, there is an issue of vocabulary. If neither the analytic continuum nor the physical one are part of the physical realm, why do we refer to the latter as physical?

The answer is as follows. Neither π nor Sherlock Holmes are part of the physical realm, but—as illustrated by the comic strip in Figure 3.2—Holmes smokes a pipe and plays a violin, which are qualities of that realm;

---

[3] *see also*: Slawinski (2016, Section 4.2)

π exhibits neither of those qualities. Mechanical continua and models of the Earth exhibit qualities of the physical realm, such as mass and rigidity.

Also, if we set coordinates on the geometric, mechanical or physical continuum, we can select a point and say that it is $\sqrt{2}$. Strictly speaking, this is false. We should refer to that point by relating it to the analytic continuum. To speak in such a manner would be pedantic, and we do not advocate it. However—if we are concerned with foundations—it is important to know the correct expression. Herein, the analytic continuum is used as a model for the geometric, mechanical and physical continua.

3.2 Continuum mechanics: Primitive concepts, first principles

3.2.1 *Predictions, observations, justifications*

Seismological theory is based on continuum mechanics and cannot exist in its present form without the foundations provided by the concept of the physical continuum. Waves and rays are intrinsic parts of differential equations, which, in turn, require the mathematical continuum. Continuum mechanics itself is not based explicitly on another theory; it is a basic—not a based—theory. By adopting it as a framework, seismology becomes an idealization, which is of crucial importance in understanding its conceptual foundation.

Continuum mechanics can be considered as a formal theory of interest in its own right regardless of its being or not being a true, or even approximately true, description of nature. In this regard it is similar to Euclidean geometry, whose interest transcends its correctness as an account of the spacetime we inhabit. The theory starts from a set of hypotheses and proceeds deductively. Formally, the conclusions drawn from the theory are of no greater generality nor are they better supported than the premises.

Even though, continuum mechanics—in its physical applications, for which it provides a realm of abstract models and analogies—deals with bulk matter of our immediate experience, its physical justification comes *a posteriori* by comparing its predictions to observations. We take continuum mechanics as a model for, not as a literal description of, nature. Similarly, if we refer to the Italian cycling champion, Fausto Coppi, as *Airone*, which means heron, we do not describe him literally, as illustrated in Figure 3.3, but we refer to his appearance and racing style by means of the metaphor.

We are aware that a hypotheticodeductive formulation does not mean that false premises must entail false conclusions. Observable predictions

Fig. 3.3

can support a theory, but do not guarantee its correctness. Consider a theory containing two principles.

1. Anything containing arsenic is nutritious.
2. Broccoli contains arsenic.

This theory implies that broccoli is nutritious, which is true, and can be tested experimentally, even though both principles are false.

The fact that a false theory can have true observable consequences is the basis for many philosophical problems about inductive inference. Also, a hypothesis may have probabilistic consequences as well as deductive ones.[4]

Although continuum mechanics is selfcontained and does not require foundations from another theory, it may still be explained in terms of other theories. Condensed-matter physics and quantum electrodynamics, which consider microscales of matter, are possibilities for such explanations. Condensed-matter physics cannot capture continuum mechanics exactly, but, in principle, can provide an approximation that is empirically

---

[4]As commented by Nola and Sankey (2007, p. 170), a more descriptive name for a hypotheticodeductive formulation might be a hypotheticoinferential method.

correct (Aoki *et al.*, 2000); and so can quantum electrodynamics, as suggested by Richard Feynman (1985/2006).

> The idea that light goes in a straight line is a convenient approximation to describe what happens in the world that is familiar to us; it is similar to the crude approximation that says when light reflects off a mirror, the angle of incidence is equal to the angle of reflection.

The attitude expressed here is common, but it might be puzzling that Feynman speaks of a crude approximation. In the world that is familiar to us, light may travel in a straight line. Also, how could standard quantum electrodynamics approximate that, given that nothing can have a trajectory in quantum theory, since that would violate the uncertainty principle? Feynman's version of quantum electrodynamics involves a sum over paths, a process that assumes a photon traverses every possible path between two points.[5] In neither of these approaches can quantum electrodynamics be considered a proper approximation to our familiar account of light.

There is another sense of approximation that Feynman might have in mind. In this other sense, we only concern ourselves with what can be observed. Theories approximate one another if empirically they are equivalent to one another, which means that they imply the same observations. In this sense, it may turn out that condensed-matter physics and quantum electrodynamics make the same empirical claims as continuum mechanics, at least when dealing with seismic phenomena.

If we think of those theories as descriptively true—unlike continuum mechanics, which we take to be a significant idealization—we may abandon continuum mechanics in their favour. However, there is no such prospect in the foreseeable future. Furthermore, even if there is a successful alternative, continuum mechanics, like classical mechanics, might remain easier to use and to generate intuitive insights. Thus, continuum mechanics is likely to remain a fundamental tool of seismology.

3.2.2   *Logical structure of theory*

Presenting a theory in its logically ideal form, we begin with undefined terms, known as primitive concepts. Other concepts are defined in terms of these primitives. Thus, set theory begins with set and its member. Other concepts such as subset, function and ordinal number are defined

---

[5] *see also*: Slawinski (2016, Sections 7.2.6 and 7.2.7)

in terms of the initial concepts. In classical mechanics, space, time and mass are primitives. Velocity is defined in terms of space and time, and momentum in terms of velocity and mass. This might appear obvious, but when conceptual and foundational issues are foremost, care must be taken.

Any attempt to define primitive concepts of a physical theory would result either in a circle, A means B, B means C, C means A, or in an infinite regress, A means B, B means C, C means D, and so on. Often, however, primitives can be understood by pointing to examples or analogies. Such an explanation might be insightful, but is not part of the formal presentation of the theory.

In a physical theory, primitive concepts should be contained in at least one physical law, which is a way of indirectly assigning them a meaning. These concepts and the laws involving them cannot be proven. Laws and the concepts they employ are postulated as axioms. A theory is a conjecture to be evaluated indirectly, usually by means of its empirical consequences. Empirical tests and other considerations mentioned in Chapter 2 provide methods of evaluation.

The process is complex and subtle. How terms acquire their meaning remains an interesting subject. David Hilbert, discussing the foundations of geometry, claimed that terms are implicitly defined by the axioms in which they occur. Gottlob Frege disagreed; according to him we need to understand the basic terms first, then say something true about them in the axioms. He pointed out that if terms are defined by the axioms, then a change of axiom would be a change of meaning of the term. Thus, point would mean something different in Euclidean and non-Euclidean geometry, which Frege took to be absurd. Hilbert, however, was quite prepared to accept this consequence.[6]

The problem of meaning has arisen recently in numerous contexts. Thomas Kuhn (1996) claimed that a term such as mass means different things in classical and relativistic mechanics. This is part of his doctrine of incommensurability, discussed in Section 2.1.4 and, in particular, in Section 2.1.4.3. One of the troubling consequences of Kuhn's view is that theories cannot contradict one another. If Einstein says that mass varies with speed and Newton says that it does not, then they are talking about different things, so—at least from a logical point of view—they no more contradict one another than the statement that apples are red contradicts that bananas are not red. There is still no fundamental consensus on the issue of meaning.

---

[6] Readers interested in philosophy of mathematics might refer to Brown (2008, Chapter 7).

Modern foundations of continuum mechanics result from the work of Clifford Truesdell and his associates. Stemming from their work, continuum mechanics metamorphosed from a bag of tricks into an elegant theory.[7] As a logical structure, the theory begins with three primitive concepts upon which it is formed.[8]

3.2.3 *Material body*

The first primitive concept of continuum mechanics is the material body that possesses the following properties. Every sufficiently smooth portion of a body is a body, and the mass of a body is the sum of the masses of its parts. The mass of a material body occupying a certain volume is represented mathematically by the volume integral of mass density. The material body itself is not accessible to direct observations. It is an abstract entity whose representations are encountered in particular spatial locations and times.

Formally, a material body, $\mathcal{B}$, is represented by a Euclidean three-dimensional smooth manifold. There is a measure called mass, m, which is a nonnegative scalar quantity such that

$$m\left(\mathcal{B}_1 \cup \mathcal{B}_2\right) = m\left(\mathcal{B}_1\right) + m\left(\mathcal{B}_2\right),$$

where $\mathcal{B}_1$ and $\mathcal{B}_2$ are disjoint subsets of $\mathcal{B}$. Specifically, the mass of a material body occupying volume V is

$$m\left(\mathcal{B}\right) = \int_V \rho \, \mathrm{d}V, \qquad (3.2)$$

where ρ is the mass density of the material composing $\mathcal{B}$. Herein, ρ is a positive locally integrable function, which implies that a compact body has a finite mass, and $\mathrm{d}V$ is the Lebesgue measure[9] of the Euclidean space, $\mathbb{E}^3$. $\mathbb{E}^3$ is a three-dimensional geometric continuum, introduced in Section 3.1. It is based on the analytic continuum, $\mathbb{R}^3$, with further properties, such as the Euclidean metric to define distances and angles.

Herein, $m(\mathcal{B})$ is referred to as the mass of body $\mathcal{B}$, even though $\mathcal{B}$ is a manifold, which is an abstract entity akin to a multidimensional surface.

---

[7]Readers interested in history of the development of continuum mechanics might refer to Maugin (2013). This work contains an extensive bibliography at the end of each chapter.

[8]*see also*: Slawinski (2015, Section 2.1) and Slawinski (2016, Sections 7.2.6 and 7.2.7)

[9]Readers interested in the Lebesgue integration might refer to Carter and van Brunt (2000).

This issue is addressed in Section 3.1, above, and discussed further in Section 3.3, below. Strictly speaking, $m(\mathcal{B})$ is not the mass of a body nor is it the mass of a manifold; it is the number that represents a mass.

3.2.4 *Euclidean spacetime*

The second primitive concept of continuum mechanics is the Euclidean spacetime. This is a generic concept, which appears in other theories, and belongs to nonrelativistic protophysics, which is the common background to many physical sciences.[10]

Euclidean spacetime is $\mathbb{E}^3 \times \mathbb{R}$, where $\mathbb{E}^3$ is composed of points $\mathbf{x}$ and $\mathbb{R}$ refers to time. Symbol $\times$ stands for direct product; hence, $\mathbb{E}^3 \times \mathbb{R}$ is the set of all ordered pairs $(\mathbf{x}, t)$.

Remark 3.1. There are two ways of understanding the notion of a protophysical entity. In one sense, it is a concept that is essential for all science. Space and time are plausible examples, since they seem required by all sciences. In the other sense, a protophysical concept is any concept incorporated without explanation. It is taken as understood due to its being established elsewhere. Thus, spacetime can be adopted by a seismological theory, since it is justified by its use elsewhere in physics.

If it turns out that spacetime is discrete, then classical continuum mechanics is doubly false. It treats atomistic matter as continuous and it assumes that it moves in continuous spacetime, $\mathbb{E}^3 \times \mathbb{R}$. In spite of this, we might use continuum mechanics to describe seismic processes and construct a theory of the Earth's interior.

3.2.5 *Stress vector*

The third primitive concept of continuum mechanics deals with the system of internal forces within the body by defining the concept of stress, which we associate with the stress vector, also known as traction,

$$\mathbf{T} := \frac{\Delta \mathbf{F}}{\Delta S}, \qquad (3.3)$$

as ΔS tends to zero.[11] ΔS is the element of surface area to which force $\Delta \mathbf{F}$ is applied.

---

[10] *see also*: Slawinski (2016, Section 2.2.1)
[11] *see also*: Slawinski (2015, Sections 2.3 and 2.5)

The magnitude of **T** remains finite as the element of the surface area becomes infinitesimal. This property, proven by Cauchy, is necessary to formulate the theory of continuum mechanics. If **T** tended to infinity, we could not formulate a physically meaningful theory that describes stress at any point within the continuum.

Not allowing infinity might be a stipulation whose aim is to avoid singularities in the equations that represent physical laws. As such, it might result from our ignorance of mathematical alternatives to represent infinite forces or, perhaps, there are no infinite forces, amplitudes and velocities in physics. Einstein and Infeld (1938) claim that

> infinite speed cannot mean much to any reasonable person.

Presently, we do not accept infinite velocities because of special relativity, but quantum nonlocality is at odds with that statement.

An example of an infinite quantity that results from a mathematical formulation, as opposed to a physical situation, is the concept of caustics in ray theory, where the amplitude of a signal becomes infinite.

Example 3.1. Caustics

A common example of a caustic is a focal point in geometrical optics. We can visualize an infinite number of rays converging at that point, where—as a consequence of the mathematical formulation—the superposition of the signal amplitudes tends to infinity.

In physics, the amplitude increases significantly at a focal point but it does not become infinite; the word 'caustic' refers to burning caused by the intensity of light. Also, there is an alternative formulation of wave theory that does not result in infinite amplitudes.

The behaviour of a signal traversing a caustic has been a challenge of modern mathematics in the context of singularities, with the so-called catastrophe theory playing an important role therein (Arnold, 1992; Berry and Upstill, 1980; Nye, 1999; Thom, 1993), whose ideas—at least partially—originate in the work of Christiaan Huygens in the seventeenth century (Porteous, 1994). Philosophically, caustics and the catastrophe theory are discussed by Batterman (2002, Section 6.3).

3.2.6   *Stress tensor and stress principle*

Let us return to continuum mechanics. Unlike the primitive concept of spacetime, the primitive concept of stress vector belongs only to continuum mechanics, and is central to this theory. It leads to the physical law known as Cauchy's stress principle.[12]

> The action of the continuum exterior to the closed surface upon the continuum within this surface is represented by a vector field, $\mathbf{T}$. The magnitude of $\mathbf{T}$ depends solely and linearly on the orientation of this surface given by its unit normal, $\mathbf{n}$.

This law, which is concisely expressed as $\mathbf{T} = \sigma\,\mathbf{n}$, and in terms of components can be written as

$$T_i = \sum_{j=1}^{3} \sigma_{ij} n_j\,, \qquad i = \{1, 2, 3\}\,,$$

is assumed to be valid for all materials considered in the realm of continuum mechanics. Herein, σ is the stress tensor, which linearly relates the vector field to the orientation of the surface. This law leads to an internally consistent and empirically adequate theory.

Given the primitive concepts, the linearity of the relation between the vector field to the orientation of the surface is entailed as a theorem. This is different from the linearity of Hooke's law, discussed in detail in Chapter 5, which is our choice based on a satisfactory level of accuracy provided by that law. We could make this law nonlinear if greater accuracy is required.

3.2.7 *Assumptions and axioms*

3.2.7.1 *Assumptions*

If one is motivated only by observational accuracy, one could justify ignoring nonobservable features within a theory. If one seeks a description of reality, one takes observational accuracy as evidence, but not as an end in itself. In such a case, an assumption, such as linearization, might become questionable. If we seek the truth, then a more accurate description of the unobserved should be our aim, even if it makes no difference to the observations we currently make.

Our response is stated on page 5, above, and discussed on page 74, below. The theory of seismology does not begin with continuum mechanics, it begins with the assertion of a granular Earth, which we take to be true. In passing, we note that even this starting point could be viewed as an idealization in which granularity is a result of elementary-particle physics. In spite of accepted granularity, we assert that the Earth could be

---

[12]*see also*: Slawinski (2015, Section 1.2.2)

modelled by a continuum. This claim is not a statement about seismological theory; it is part of that theory. A geophysicist would agree that the continuum-mechanics quantities, such as mass density, anisotropy, rigidity, carry information about discrete and granular properties of the Earth, such as fractures, porosity, layering, crystals. The above distinction is important both for applying seismological theory and for examining its foundations.

This point arises when considering details of any theory. In most cases, including nonlinearity within a theory achieves greater accuracy. However, commonly, information about nonlinearity is ignored in the context of examined phenomena. An empiricist would choose such an approach because there is no empirical difference between the linear and nonlinear formulations. A realist would ignore nonlinearity as an idealization of reality, whose generalizations are not observationally detectable.

To clarify, let us consider the differences among three planetary theories: geocentric, heliocentric with circular orbits and heliocentric with elliptical orbits. Consider them as observationally equivalent, as would have been the case in the late sixteenth century. An antirealist would be indifferent about the choice of the theory. A realist might say that the elliptical version is the truth, but might adopt the circular orbits, and justify this as a reasonable idealization due to its convenience. This could not be said of the geocentric theory, since it is not an idealization of the heliocentric theory, even though it is observationally the same.

3.2.7.2 *Axioms*

The purpose of primitive concepts embedded in the axioms of continuum mechanics is to establish a theory that describes both observable phenomena and the unobservable realm. The axioms of this theory are neither selfevident truths nor arbitrary formal stipulations. They are conjectures. Consequently, they are fallible, and subject to revision.

Broadly speaking, there are three views concerning axioms. The first view is that they are selfevident truths. This is a view commonly associated with the history of Euclidean geometry. Euclid's postulates were taken to be obvious and certain. Their consequences, thanks to the infallibility of deductive logic, were equally certain.

The second view is that axioms are arbitrary. Some mathematicians, known as conventionalists or formalists, hold this view about mathematics, and say that we have complete freedom in postulating whatever we want provided we uphold consistency. This view is favoured by those who think of mathematics as a game, played with arbitrary rules that we could change, just as we could change the rules of chess. Various postulates might be more or less interesting and useful, but none are, strictly speaking, true.

The third view is that axioms are fallible attempts to describe how things are. Like the first option, it is a realist view, but claims no certainty. Axioms are conjectures tested by their consequences, which include everything from direct observations to the ability of a theory to explain a wide variety of phenomena and solve outstanding problems. In adopting continuum mechanics in seismology, we accept this third view.

Attitudes towards axioms are of importance within debates over the nature of mathematics, where each of the three views has its champions. Though less apparent in the natural sciences, similar differences of opinion manifest themselves. Einstein distinguished principle from constructive theories. The former he took to be more or less certain, while the latter he took to be speculative conjectures. Special relativity's principle of relativity and the constancy-of-light-velocity postulate were, for him, more or less certain, and used as constraints on everything else. The molecular-kinetic theory of matter and heat, on the other hand, was for him a conjecture, to be tested by its ability to explain and predict. The attitude of many to classical mechanics is, in important respects, similar to the formalist view of mathematics. Its actual truth is—currently, not historically—of little or no concern. One is only interested in new and remarkable consequences of the theory, such as deterministic chaos.

In seismology, or any other natural science, there are grades of difference and interactions among the three broad views. A model might be adopted in the spirit of a conjecture, but nevertheless function in an *a priori* way. It is taken to be certain and inviolable while other assumptions are modified to account for observational details.

A standard approach of seismology is to begin with postulating a background model, such as the spherical symmetry of layers within the Earth, on a global scale, or lateral homogeneity of layers within sedimentary basins, on a regional scale. A background model might not be based on seismic data. It is an assumption or conjecture, which might be suggested by information from planetary physics and geology.

Having accepted a background model, a seismologist attempts to refine it using the available data. A commonly used term of 'seismic anomaly' and a ubiquitously used mathematical technique of the Fréchet derivative, attest to such an approach. The former refers to the discrepancy between the computed observables based on the background model and the measured ones. The latter one is a quantitative technique for such a discrepancy, commonly referred to as a misfit.

Another part of the standard approach is to consider a particular continuum that is defined by a Hookean solid. Such a solid might begin as a conjecture but then operate as an *a priori* framework, and hence become an unmodifiable constraint.

At what point does the *a priori* framework give way to another? To evaluate such a framework, one must enquire if the discrepancy between the computed results and the observations might be accommodated by refining the background model, while maintaining its general structure. One might study how changing the background model would affect the predictions, and how plausible are these alternative models.

Given that axioms are conjectures, we face a problem of evidence. As discussed on page 57, a physical theory might be false yet lead to formulations that result in satisfactorily accurate predictions. Imagine a theory according to which we live in the interior of a hollow sphere. It follows that if we travel in any direction we return to our starting point. Empirically, we find this is so, even though we live on the outside of a sphere. In short, false theories can have true testable consequences. In general, testable consequence does not exhaust the notion of evidence.

Classical mechanics is an example of a theory based on assumptions of limited physical validity. For instance, the assumption of conservation of mass is limited to cases in which speeds are small compared to that of light. A modification of the concept of mass is proposed and justified by the theory of relativity.[13]

Until the formulation of the theory of relativity, conservation of mass was viewed as a protophysical principle. The notion of a continuum in the context of seismology never was. It has been viewed as a formalism that results in sufficiently accurate predictions. In this sense, the assumption of a continuum has aspects of a formal stipulation whose only requirement is a consistent—which means, not leading to contradictions—use in the subsequent derivations.

---

[13]Readers interested in the relationship between Newton's second law of motion and special relativity might refer to Koks (2006, Section 5.9).

3.3 Applied mathematics

For the foundations of seismology, we assume two distinct realms: abstract and physical, which is a Platonic approach, whose essence is shown in Figure A.1, on page 129. The physical realm contains the Earth, and the abstract realm is rich enough to represent the physical one. Thus, to examine an aspect of the physical world, we find a mathematical structure to represent it.

Platonism is common approach among mathematicians and mathematical physicists (Penrose, 1997, Figures 1.1–1.3), (Penrose, 2004, Chapter 1). Also, it is endorsed by many philosophers of mathematics.[14]

Example 3.2. Weights and nonnegative real numbers
Weight is represented on a numerical scale. The main physical relations among objects that have weight are that some have more weight than others and that if combined, their joint weight is greater than either of their individual weights. Weight can be represented by a mathematical structure of the nonnegative real numbers, in which there is a greater-than relation matching the physical greater-than relation, and an addition relation matching the physical combination relation.

More generally, a mathematical representation of a nonmathematical realm occurs if there is a homomorphism, ϕ, between a physical system, $\mathbf{P}$, and a mathematical system, $\mathbf{M}$. $\mathbf{P}$ consists of a domain, D, and relations, $R_1, R_2, \ldots$, defined on that domain; similarly, $\mathbf{M}$ consists of D^* and $R_1^*, R_2^*, \ldots$. A homomorphism is a mapping from D to D^* that preserves the structure in the appropriate manner.

Example 3.3. Homomorphism of weights and numbers
Continuing with Example 3.2, we let D be a set of bodies with weight, and $D^* = \mathbb{R}_+$, be the set of nonnegative real numbers. Furthermore, we let $\preceq$ and $\oplus$ be the relations of, respectively, *physically weighs the same or less than* and *combination of physical weights*. Relations $\leq$ and $+$ are the familiar relations on real numbers of *equal to or less than* and *addition*. The two systems are $\mathbf{P} = \langle D, \preceq, \oplus, u \rangle$ and $\mathbf{M} = \langle \mathbb{R}_+, \leq, +, 1 \rangle$. Numbers are associated with the bodies, $\{a, b, \ldots\} \in D$, by the homomorphism, ϕ:

---

[14]Readers interested in arguments for and against Platonism in mathematics might refer to Brown (2008), Colyvan (2001) Kitcher (1983), Leng (2010) and Shapiro (2000).

$D \to D^*$, that satisfies three conditions.

$$a \preceq b \to \phi(a) \leq \phi(b),$$
$$\phi(a \oplus b) = \phi(a) + \phi(b),$$
$$\phi(u) = 1.$$

The first expression says that if a weighs the same or less than b, then the real number associated with a is equal to or less than the real number associated with b. The second expression says that the number associated with the weight of the combined object $a \oplus b$ is equal to the sum of the numbers associated with these objects separately. The relations that hold among physical bodies get encoded into the mathematical realm and are represented there by relations among real numbers. One of the objects, u, can be singled out to serve as the unit weight, so that $\phi(u) = 1$. The unit is usually given a name, such as kilogram.

The crucial point is that mathematics applies to the physical world by providing models, which are analogies to structures and processes in the physical world. Mathematics is not a direct description of that world. Mass is not a number and force is not a vector, but they are represented by real numbers and vectors, respectively. The common phrase of a mathematical description of reality is misleading if it means anything more than mathematics being used as an abstract model of, or an analogy for, physical phenomena.[15]

Measurement theory classifies different types of scale. The simplest of these are ordinal measurements. For example, the Mohs scale of hardness uses integers from 1 to 10 in ranking the physical relation of objects scratching each other. Talc is 1 and diamond is 10. Addition plays no role; the only property of the numbers used is their order—a strict order: nothing scratches itself. There are no metrical assumptions; that is, there is no sense of distance between two places on the scale. Interval measurements use the greater-than relation between real numbers, but do not employ addition. Temperature is another example of such a structure. Two bodies at 50° each do not combine to make one at 100°; there is, however, a formula—akin to averaging—for combining the temperatures of two bodies. By contrast, simple addition is used in extensive measurements, such as measurements of weight.

The physical combination of two bodies with weights, as illustrated in Examples 3.2 and 3.3, is represented by the addition of two real numbers.

---

[15] Readers interested in philosophy of applied mathematics might refer to Brown (2008, Chapter 4).

However, as alluded in reference to temperature, the embedding homomorphism is not always as simple as with weights. The relativistic addition of two velocities, for example, is given by an expression constrained by the upper limit, which is the velocity of light (Brown, 2008, p. 61).

The mathematical representation of the world need not be with numbers. From the Greeks to Galileo, geometrical objects did the representing. The increasing speed of a falling body, for example, was represented by Galileo as a sequence of increasing areas of geometrical figures. Also, Newton's *Principia* was written in this geometrical style. Thereafter, the power of calculus has made analysis dominant. The geometric spirit, however, persists, as exemplified by the visual book of Abraham and Shaw (1992).

There are further subtleties. Diagrams and graphs are geometrical, but they can depict numerical results, and be geometrical representations of analytic representations of the physical realm. Also, analytic entities have many representations; for instance, functions can be given by formulæ, graphs or tables of values.

There are differences in the degree of abstractness among mathematical models. Attaching numbers—that is, counting—is perhaps the most transparent application. Associating speed with a derivative is more abstract. Assigning probabilities to dice rolling is akin to counting, but assigning Hilbert spaces and methods of calculating probabilities in quantum mechanics is more sophisticated. Yet, they are all models.

3.4 Models, fictions and idealizations

3.4.1 *Analogies*

Models play a role in analogical reasoning. For instance, Bohr's model of the atom is based on an analogy of the Sun and orbiting planets with the nucleus and orbiting electrons. The literature on models is vast.[16]

Prior to discussing different aspects of models, we need to deal with the fact that the term model has many uses. One of these is tentative, as when a scientist says that a given idea is not a theory yet, since there is not enough evidence, but is only a model. This is not the sense of model that interests us here. Furthermore, we are not concerned with the sense of model from logic, where a model is an abstract structure that serves as an

---

[16]Readers interested in discussions on models might refer to Giere (2004), Morrison (2015), van Fraassen (2008) and Weisberg (2013).

interpretation of a formal theory, making the sentences of that theory true in the interpretation. Nevertheless, there is an overlap with both of these, that is, with the idea of tentativeness and with the notion of interpretation of a formal system.

3.4.2   *Physical models*

Physical models are material objects. A spring can be a physical model for a Hookean solid, as illustrated playfully in Figure 5.1 on page 90.

Physical models are commonly used in seismology. They might consist of laminar materials, composed of rocks or other substances, that—together with transducers and receivers—serve as physical models for propagation of disturbances in sedimentary layers. They serve experimental purposes of understanding the effects of geological structures on propagation of recorded disturbances, just as a model of a plane tested in a wind tunnel serves in adjusting an aeronautical design. In both cases, the issue of scale needs to be addressed to ensure that the behaviour in the laboratory setup is an adequate model for a sedimentary basin or for a full-size plane.

These models work in the same manner as do mathematical models, discussed in Section 3.4.4, below. The models have targets, such as sedimentary layers, airplanes, and human cancer. We subject them to various procedures such as examination of propagation of mechanical disturbances in a laminar material and of a plane model in a wind tunnel. Various measurements are made, such as reflections of a signal within layers and frictional force on the model plane. Inferences are subsequently drawn about the targets of the models, such as the seismic effects of a fault in a layered medium or an adjustment of the wing tips to increase stability of a plane. Whether we are manipulating and measuring a physical model or altering and calculating a mathematical model, the pattern is the same. We form a link between the object of interest and its model to make inferences about that object. Even though the underlying pattern is straightforward, challenges might present themselves when considering further details.

We require a model to exhibit a relevant similarity. For example, in pharmacological studies, a mouse is similar to a human with regard to aspects of its metabolism, but not its tail or its general appearance, as illustrated playfully in Figure 3.4.

Fig. 3.4

3.4.3 *Data models*

To examine observations, we consider models of data. The importance of such models was emphasized by Suppes (1962). Its various aspects are examined by Bogan and Woodward (1945), Cartwright (1994), McAllister (2007).

Curve fitting is a typical example of data modelling. The process depends on the criterion of the fit, which is the choice of a scientist. Commonly, it could be the least-square fit or the least-absolute-value fit.

However, scientists rarely work with raw data. They construct data from observations in a manner that is conditioned by background assumptions and even by the theory they are to test. For instance, a seismograph might record the intensity of an electric current induced by the motion of the magnet within it. However, it is assumed that this motion is due to a disturbance caused by a distant quake and propagating through the Earth.

3.4.4 *Mathematical models*

Mathematical models are essential for theoretical seismology as the tools for prediction and retrodiction of observed phenomena, as well as for the inference of terrestrial properties. However, as stated by Fowler (1997, p. 3),

> [m]athematical modeling is a subject that is difficult to teach. It
> is what applied mathematics (or, to be precise, physical applied

mathematics) is all about, and yet there are few texts that approach the subject in a serious way. Partly, this is because one learns it by practice: there are no set rules, and an understanding of the 'right' way to model can only be reached by familiarity with a wealth of examples.

Mathematical models, as the name suggests, are based on mathematical reasoning. As examples, consider a mass bouncing on a spring, a pendulum and an electric circuit. They are modelled by the same differential equation,

$$\frac{d^2 x}{dt^2} = Ax, \qquad (3.4)$$

where t stands for time, and x and A assume different meanings in different contexts.

For a mass on a spring, x stands for displacement and $A = -k/m$, where m stands for mass and k is a spring constant, which is a parameter quantifying the stiffness of the spring. For a pendulum, x stands for the angular displacement, which is assumed to be small, and $A = -g/\ell$, where g stands for acceleration due to gravity and ℓ for the length of the pendulum. For an electric circuit, x stands for the charge and A^{-1} is the negative product of capacitance and inductance.

In view of obeying the same differential equation, different physical processes are analogous to each other and can serve as models for each other. Notably, a significant part of seismological theory uses the results of mathematical models created originally for the study of light propagation, since the electromagnetic and mechanical waves exhibit aspects of a formal analogy.

3.4.5 *Fictions and idealizations*

Fictions and idealizations are of importance in seismology. The frictionless plane is a key example of both. It does not exist in the physical world; hence it is a fiction. However, it can be approximated by progressively smoother surfaces; hence, it is an idealization. A homogeneous layer used commonly in seismology is also a fiction. A geologist could protest that there are many intricacies within a layer that are ignored by the assumption of homogeneity.

In philosophical discussions, we distinguish between two idealizations, namely, Aristotelian and Galilean (McMullin, 1985). An Aristotelian idealization ignores various properties that are considered irrelevant. For instance, discussing gravity, the colour of a falling body is ignored.

A Galilean idealization reaches further. For instance, discussing gravity, a material body is treated as a point. Similarly, the granular Earth is treated as continuous.

An Aristotelian idealization never states anything false. Galilean idealization does. The philosophical question is simple but profound. Why does Galilean idealization work in spite of false statements?

Alexandre Koyré (1968) states that

> The scientific revolution happened when people stopped looking and started thinking.

Herein, thinking does not refer directly to a falsity of an idealization, but to thought experiments, which also rely on Galilean idealizations. For instance, Galilean relativity is the result of a thought experiment involving being shut inside a ship that might move at a constant velocity or be at rest in the harbour to argue that we are incapable of distinguishing between the two cases by mechanical experiments performed within the ship.

3.5 Seismological continuum

Heuristically, and in contrast to the presentation in Section 3.1, it is common to introduce the notion of a mechanical continuum in the context of particle mechanics by considering a large number of particles. For instance, considering a pile of sand, one might consider average properties such as mass density and compressibility of the entire pile, not paying attention to individual grains. In geology, one implicitly invokes an aspect of a continuum when referring to the density of granite, since it is an average that combines minerals of quartz, mica and feldspar.

Seismological justification of the applicability of a mechanical continuum is consistent with such a heuristic description. Bulk matter is immediate to our experience and so are its properties. We experience viscosity and incompressibility of water, not the properties of individual dihydrogen-oxide molecules. However, from the foundational viewpoint, continuum and particle mechanics are distinct physical theories, as discussed by Bunge (1967, pp. 143 and 155).

The continuum used to model seismic phenomena is an abstract entity that has properties similar to physical bodies. It has mass, and transmits energy and momentum. For that reason—as discussed on page 55, using an illustration of π and Sherlock Holmes—the seismological continuum is distinct from a purely mathematical entity. In this book, the seismological continuum is equivalent to the physical one, which is discussed in Section 3.1. Also, it is tantamount to geophysical continua discussed by Kennett and Bunge (2008), and used as the title of their book. This continuum is a model of the Earth, and it is in turn modelled mathematically by $\mathbb{E}^3$, as stated on page 61.

Let us postulate the first principle of seismological theory.

> I. Earth is a granular body, intrinsically composed of discrete particles, but—for the purposes of seismology—assumed to be a physical continuum.

The evidence that supports this assumption is based on macroscopic observations. Furthermore, we assume the unobserved interior of the Earth to exhibit the same properties. By granular—*sensu lato*—we mean discrete structures, which are represented within a continuum by its anisotropy and inhomogeneity. Rotational and translational symmetries carry the information about these structures.

We postulate the second principle of seismology to be the following.

> II. Continuum mechanics is the quantitative framework for seismology.

In other words, the formulation of continuum-mechanics quantities—such as mass densites, elasticity parameters, and hence, expressions for resulting wave-propagation speeds,[17] in accordance with the balance of linear momentum—are used directly in seismological studies. On the other hand, assumptions allowing seismologists to postulate values of these quantities and to make inferences of terrestrial properties based on their results belong to the theory of seismology.

The justification for the second principle is the agreement between the observed seismic phenomena and predictions provided by continuum mechanics. There is also a pragmatic issue of formulating quantitative methods that would be more difficult or even impossible without Principle II.

Principles I and II emphasize the distinction between physical and abstract entities resulting from modelling inhomogeneous physical materials by inhomogeneous anisotropic continua (Backus, 1962; Bos *et al.*, 2016). Anisotropy, which is the variation of properties with direction, is a macroscopic property only. Its counterpart at a microscopic scale consists of different directional patterns of inhomogeneity, which is the variation of

---

[17] *see also* : Slawinski (2015, Sections 6.1.1–6.1.3, Sections 9.2.2 and 9.2.3)

properties with position. There is no anisotropy at a microscopic scale, only inhomogeneity. For instance, an anisotropy exhibited by a crystal is due to the inhomogeneity of its lattice arrangement. Thus, macroscopic properties change as functions of direction due to patterns of inhomogeneity, such as layers, grain orientations, lattice arrangements. However, in continuum mechanics, in contrast to physical materials, anisotropy and inhomogeneity—as well as their counterparts of isotropy and homogeneity—are distinct properties, as discussed by Malvern (1969, p. 2).

Remark 3.2. Seismological continua that are reasonable models of the Earth are layered. Each layer might be a distinct Hookean solid. Commonly, by invoking the assumption of a welded contact,[18] seismologists set the necessary conditions at interfaces between Hookean solids, and hence, can derive reflection and transmission coefficients[19] as well as equations of motion[20] within layered media. This assumption partially circumvents the issue of contact interaction and impenetrability within continuum mechanics, discussed by Smith (2007b).

Since velocities of waves are the seismological quantities derived from mechanical properties of continua—such as the elasticity parameters and mass density of a Hookean solid—it is convenient to describe a seismological continuum by specifying the velocities of seismic waves within it. Rogister and Slawinski (2005) formulate Hamilton's equations for a continuum whose inhomogeneity is expressed by the velocity increasing linearly with depth, and its anisotropy by the elliptical dependence of velocity with direction.[21] The mathematical model of this seismological continuum is a geometric continuum endowed with a metric to provide a measure of distances and angles.

Herein, the epistemological simplicity—discussed in Section 1.3.2, above—allows for an analytical solution of Hamilton's equations, which—given the seismological properties of a continuum, and the locations of sources and receivers—includes the explicit expression for the travel time of the seismic signal. Furthermore, as postulated by Slotnick (1959, Part IV) and examined further by Slawinski *et al.* (2004), such a formulation is empirically accurate to account for the travel time of seismic signals in

---

[18]Readers interested in the concept of contiguity within a continuous medium, in general, might refer to Hesse (2005/1961, Chapter VIII, pp. 195–196), and for the welded contact, in particular, to Chapman (2004, Section 4.3.1).
[19]*see also*: Slawinski (2016, Chapter 6)
[20]*see also*: Slawinski (2015, Chapter 10)
[21]*see also*: Bóna and Slawinski (2015, Section 3.4.1)

a sedimentary basin. Also, it is consistent with the Simplicity Postulate, proposed by Jeffreys (1931) and discussed by Anduaga (2016, Chapter 5).

Closing remarks

Concepts presented in this chapter allow us to examine theoretical aspects of continuum mechanics and its mathematical apparatus as bases of seismology. As a science based on continuum mechanics, seismology relies on Galilean idealizations, in which we invoke abstract concepts. These idealizations are connected to the Earth by the empirical adequacy of seismological predictions and retrodictions of terrestrial phenomena.

Even if seismologists choose to abandon continuum mechanics in favour of another Galilean idealization, say, condensed-matter physics, the essence of the mathematical apparatus will remain unchanged, since mathematics is the structure underlying any such quantitative theory. Mathematics, which relies on Galilean idealizations, is an analogy for models of the Earth. These models can be the mechanical continuum and its behaviour, or another model described in terms of equations.

Chapter 4

Continuum mechanics: General principles, constitutive relations

Take a spring, and on one end hang a weight of one pound. The spring lengthens, say by one inch. Now lay the spring on a horizontal table, fastening one end to the center, and leaving the weight attached to the other end. Spin the table, and adjust the angular speed until the spring again stretches exactly one inch. On seeing this demonstration in the laboratory, the freshmen [...] take it as obvious that the force exerted by the spring is again one poundal [...] What has been assumed, tacitly, is that the elastic law or constitutive equation of the spring is invariant under rotation. [...] To a person lacking this belief, the experiment measures nothing. Such a person sees two events but can assert no correlation of one with another. (As is usual in "fundamental" experiments, the main point at issue must be conceded before the experiment is started.)[1]

Clifford Ambrose Truesdell (1966)

Preliminary remarks

As suggested in the preliminary remarks of Chapter 3, seismology—unlike, for instance, electromagnetism—is not a stand-alone theory. Apart from the standard structure of logic and grammar of mathematics, seismology is built upon continuum mechanics. However, seismology is not just an application of continuum mechanics to terrestrial materials.

Let us explain the latter statement by the following analogy. Considering Newton's three laws of motion, we could view his theory of gravitation

---

[1] In interpreting this experiment, the observer implicitly invokes the equivalence between the gravitational and inertial forces, since—in classical mechanics—the centrifugal force is an inertial force.

to be their application, since it suffices to specify a force as the gravitational force and derive its consequences. Initially, Coulomb's law,[2] which governs the force between electrically charged particles, was seen as an application of Newton's laws of mechanics to a specific force. However, the development of electrodynamics showed that this law is not merely an application of another theory, but exists on its own, even though it uses the conceptual apparatus of Newton's mechanics. A similar relation exists between seismology and continuum mechanics.

We begin this chapter by discussing the general principles of continuum mechanics that are pertinent to the formulation of seismological theory. These principles originate in the balance of mass, balance of linear momentum and balance of angular momentum, which are expressions of conservation laws. There is no need to invoke explicitly other general principles, such as the ones originating in laws of thermodynamics, since they do not bring any further constraints to the discussion of Hookean solids, which are restricted to processes that do not explicitly involve changes in temperature or heat transfer. We conclude this chapter by discussing the constitutive relations, whose role is to relate the general principles to empirical results. These relations allow us to consider various models of continua. Among these models, a Hookean solid, which is the subject of Chapter 5, is a common choice in seismology.

4.1   Two-part theory

Continuum mechanics—beyond the primitive concepts and first principles introduced in Section 3.2—is composed of a general theory, which applies to all materials examined, and specific theories that relate the general theory to particular applications. For instance, the balance of linear momentum, which is a general principle, leads, within the application to fluids, to the Navier-Stokes equation and, within the application to elastic solids, to the wave equation.

Specific theories do not contradict each other, even if there are statements that are contradictory between such theories. These theories are used to examine different properties and do not share the same areas of applicability.

The limited validity of specific theories allows us to study different classes of materials or a given material in different contexts. Herein, limited

---

[2] *see also*: Bóna and Slawinski (2015, Appendix C.1.2)

validity means that the theory is empirically adequate in a restricted range of applications. It does not mean that it is literally true within that range.

In elasticity theory, a Hookean solid models a behaviour solely as a function of the amount of deformation. In fluid mechanics, a Stokesian fluid models a behaviour solely as a function of the rate of deformation. Also, the two can be combined, as exemplified by the Kelvin-Voigt and Maxwell models, mentioned in Section 7.3.[3,4]

The two-part structure of the theory of continuum mechanics is a consequence of its being based on macroscopic concepts. Concepts such as rigidity are not derived from properties of intrinsic constituents of matter, such as the strength of bonds among atoms, but are introduced through specific theories. In this regard, continuum mechanics is similar to classical thermodynamics, where temperature and pressure are introduced on their own, without relating them explicitly to molecular properties. Constitutive equations in continuum mechanics are analogous to equations of state in thermodynamics.

The necessity for existence of constitutive equations is one of several similarities between continuum mechanics and electromagnetism (Bunge, 1967). In either theory, material properties are expressed by tensors, which—in the context of these theories—are the property tensors. The property tensor that is of particular interest in our studies is the elasticity tensor, whose properties allow us to quantify—by analogy between theoretical predictions and seismic measurements—phenomena within the Earth.

The reason for specific theories can be also illustrated by the following description. In continuum mechanics, solids and liquids are subjects of specific theories due to their distinct behaviours. Solids resist change of shape and liquids do not. This is their fundamental ontology, since—without abandoning continuum mechanics—we cannot inquire into the underlying atomic structure.

Invoking continuum mechanics in the context of seismology, we assume that—in spite of its intrinsic limitations—the loss of the underlying truth of atomic matter is compensated by heuristic insights and the computational ease provided by the continuum. Furthermore, in spite of these limitations, the accuracy of resulting predictions is presently beyond most experimental data.

---

[3] *see also*: Slawinski (2015, Sections 3.2 and 3.4)

[4] Readers interested in further discussions of the Kelvin-Voigt and Maxwell models might refer to Kennett and Bunge (2008, Section 1.1.3 and Figure 1.3).

4.2 General principles

4.2.1 *Elasticity-theory requirements*

Having established, in Section 3.2, the primitive basis of continuum mechanics—whose primitive concepts are material body, spacetime and stress vector—let us consider its general principles. Since seismological theory is based on a particular branch of continuum mechanics—theory of elasticity—we focus our attention on the three principles required to study that branch. These general principles employ the three primitive concepts, as required for any scientific theory.

4.2.2 *Balance of mass*

[5]The first general principle, which is rooted in the concept of material body, is the balance of mass. It states that if no mass is created or destroyed, the time rate of change of mass within a volume is equal to the rate of mass flow through the surface bounding this volume. In other words,

$$\frac{\mathrm{d}}{\mathrm{d}t}\int_V \rho\,\mathrm{d}V = -\int_S \rho\,\mathbf{v}\cdot\mathbf{n}\,\mathrm{d}S\,, \tag{4.1}$$

where ρ stands for the mass density and S is the surface enclosing volume V. The integral on the left-hand side derives from expression (3.2), which is the mathematical expression for the material body. Equation (4.1) states that the change in the amount of mass in a volume at any instant is balanced by the mass flowing with velocity $\mathbf{v}$ through the surface that encloses this volume and whose orientation at any point is specified by the outward normal, $\mathbf{n}$. Velocity is defined in terms of the primitive concept of spacetime, which is endowed with a metric to quantify distances.

The scalar product, $\mathbf{v}\cdot\mathbf{n}$, is the measure of obliqueness between the flow direction and the surface orientation. If the flow is parallel to the surface, then $\mathbf{v}\cdot\mathbf{n} = 0$, which means there is no flow through the surface, as expected.

A word about terminology. Conservation principles are particular expressions of the corresponding balance principles. The balance implies that—for a given portion of the continuum—even though the total amount of a given quantity changes, the change is accounted for by the flow of this quantity within the continuum. If there is no flow through the boundary—which might be the case if a given portion of the continuum moves together

---

[5]*see also*: Slawinski (2015, Section 2.1)

with the flow as a water-filled balloon transported within a river—the total amount contained within a given portion remains constant, and the balance is expressed as a conservation law.

4.2.3   *Balance of linear momentum*

[6]The second principle is the balance of linear momentum, which states that the rate of change of the momentum of a given mass is equal to the sum of all the external forces acting upon it. In other words,

$$\frac{\mathrm{d}}{\mathrm{d}t} \int_V \rho \, \frac{\mathrm{d}\mathbf{u}}{\mathrm{d}t} \, \mathrm{d}V = \int_S \mathbf{T} \, \mathrm{d}S + \int_V \mathbf{f} \, \mathrm{d}V, \qquad (4.2)$$

where $\mathbf{u}$ is the displacement vector; herein, $\mathrm{d}\mathbf{u}/\mathrm{d}t$ is $\mathbf{v}$ in equation (4.1).

The right-hand side of equation (4.2) represents the sum of the surface forces given by the stress vector, $\mathbf{T}$, whose definition is given in expression (3.3), and the body forces, such as gravity, given by $\mathbf{f}$. $\mathbf{T}$ is the primitive concept that represents the traction on a surface within a continuum. Mass density, ρ, stems from the primitive concepts of material body and spacetime.

4.2.4   *Balance of angular momentum*

[7]The third principle is the balance of angular momentum, which states that the time rate of change of the angular momentum of a given mass is equal to the vector sum of the external torques acting upon it. In other words,

$$\frac{\mathrm{d}}{\mathrm{d}t} \int_V \left(\mathbf{x} \times \rho \frac{\mathrm{d}\mathbf{u}}{\mathrm{d}t} \right) \, \mathrm{d}V = \int_S (\mathbf{x} \times \mathbf{T}) \, \mathrm{d}S + \int_V (\mathbf{x} \times \mathbf{f}) \, \mathrm{d}V, \qquad (4.3)$$

where $\times$ stands for the vector product. The integrand on the left-hand side is the angular-momentum density: the vector product of the distance, $\mathbf{x}$, between a reference point and the element of the continuum with the linear-momentum density, $\rho \, \mathrm{d}\mathbf{u}/\mathrm{d}t$. Hence, the rate of change of the angular momentum of an element within the continuum is equal to the sum of torques due to the external forces acting upon this element. Herein, $\mathbf{x} \times \mathbf{T}$ and $\mathbf{x} \times \mathbf{f}$ are torque densities per unit area and unit volume, respectively.

The chosen form of the balance of angular momentum is a postulate based on practical considerations, not a fundamental physical law. For

---

[6]*see also*: Slawinski (2015, Section 2.4)
[7]*see also*: Slawinski (2015, Section 2.7)

instance, the right-hand side does not include coupled forces, whose effect is assumed to be negligible in seismology in view of the sampling wavelength being orders of magnitude larger than the size of the grains composing the materials. Hence, continua considered by seismologists need not account for the effects of rotations—expressed by coupled forces—of grains resulting from the passage of waves.[8]

4.3   Constitutive relations

4.3.1   *Specific theories*

[9]The behaviour of any elastic continuum agrees with the balance of mass and of linear and angular momenta. Other general principles, such as balance of energy, balance of electric charge, balance of magnetic flux and entropy—do not provide explicit constraints for the theory of elasticity. They are superfluous within that theory.

Given only the general principles, we cannot predict the behaviour of a material until we are given information about the relationship between physical quantities in the context of this material. For instance, how much force is required to stretch a spring to a certain length. Since the amount of force depends on the constitution of the material, such relations are called constitutive relations. Any constitutive relation is a relation between stress and other physical quantities that are appropriate for considerations of a given material. Without these relations, continuum mechanics cannot make specific predictions or offer explanations of observed phenomena.

Constitutive relations are specific to a given material. Also, they are specific to a particular experimental or observational setting.

Prior to discussing these relations, let us comment on the distinction between experiments and observations, which also appears in the quote on page 87, below. This distinction might or might not affect the choice of a constitutive relation, but it is important to be aware of it while considering theoretical formulations and empirical examinations. Both processes appear in seismology. Commonly, the acquisition of data to infer information about the deep interior of the Earth relies on recording earthquake disturbances. For exploration seismology, such disturbances can be generated by

---

[8]Readers interested in the effects of microrotations, which are modelled by the media formulated by Cosserat and Cosserat (1909), might refer to Forest (2006) and to Truesdell (1966, Lecture II).

[9]*see also*: Slawinski (2015, Section 3.1)

artificial sources. Hence, the former can be viewed as an observation and the latter as an experiment.

Let us return to constitutive relations. Different materials are represented by different constitutive relations and even the same material might be represented by different relations depending on the scientific interest of the investigator. Such a relation must be a suitable idealization. Geophysicists concerned with postglacial rebound—which is a process whose duration takes millennia—model the Earth as a viscoelastic continuum. Such models allows them to include time in the process of deformation. Geophysicists concerned with earthquakes model the Earth as an elastic medium, hence allowing an instantaneous response to stresses.

At first sight, the use of different, and even contradictory, models to study the Earth presents us with a question if they can be accounts of reality. However, models capture only part of their target, which is a given terrestrial phenomenon, and do so only approximately. Distinct models focus on distinct phenomena and properties. Water cannot be both compressible and incompressible, but, in hydrology, a model of compressibility is used to study sound propagation and a model of incompressibility for tracking the water flow in a pipe. If we can trace the relation between a given model and its target, we can avoid contradicting ourselves and we can maintain a realist outlook.

The general principles are the framework for a plethora of distinct specific theories, such as elasticity, viscoelasticity, plasticity, viscoplasticity. They independently attempt to model solids, liquids, viscous and plastic materials, and so on. In spite of this variety, all constitutive relations must obey—in addition to the general principles of continuum mechanics—certain conditions, which we discuss in Sections 4.3.2–4.3.5, below.

Constitutive equations do not stem from fundamental principles, but they complement these principles. They can constrain applicability of fundamental principles—as is the case of Hooke's law restricting the applicability of the balance of momentum to elastic continua, and hence, entailing the wave equation from the balance principle—but they cannot contradict such principles.

Commonly, constitutive equations stem from idealized experimental results, such as the extension of a string being linearly proportional to the applied force, which inspired Robert Hooke to postulate his law in one dimension. In principle, however, constitutive equations could be postulated without any prior empirical motivation, and subsequently tested by their consequences. Yet, even for a latter case, one could argue that, at least indirectly, there always is an empirical motivation.

Thus, constitutive equations allow us to test empirically a theory. They are the link between the fundamental principles and experimental measurements. A seismological theory provides a structure that relates observables to the parameters that define a given continuum. Hence, we postulate the role of constitutive equations as the third principle of seismology.

III. Constitutive equations of continuum mechanics specify properties of a material within a given model of the Earth.

To specify properties of a given material, seismologists might invoke a Hookean solid or another constitutive equation. Their choice depends partially on the actual material to be studied and partially on the interests and goals of investigation. Any model, however, should obey the following three principles[10] and the requirement of equipresence. The status of each principle and of equipresence is debatable in the context of foundations of physics. However, be that as it may, they are fruitful guiding rules to construct specific theories within continuum mechanics.

4.3.2   *Determinism*

The principle of determinism, which is a tenet of classical physics, states that the stress within a body is determined uniquely by its past deformations. This also means that only past and present can affect the present state.

Hooke's law, in particular, is concerned only with the present: stress at a particular instant depends only on deformation at that instant. The behaviour of a Hookean solid is not affected by its history; such a solid has no memory of past deformations.

4.3.3   *Local action*

The principle of local action states that deformations outside an arbitrarily small neighbourhood of a point may be disregarded in determining stresses at that point. This implies that there is no action-at-a-distance between stress and strain. In other words, there is no instantaneous interaction between two points that are separated in space.

---

[10]Readers interested in these principles might refer to Truesdell (1966, Lecture I, pp. 4–6).

4.3.4 *Material frame indifference*

The principle of material frame indifference states that any two observers of deformation of a body find the same stress tensor. In other words, constitutive equations must be invariant under changes of a reference frame that preserve the essential structure of spacetime. The quote that introduces this chapter is an example of material frame indifference (Truesdell, 1966, p. 6).

4.3.5 *Equipresence*

For theories that contain more than one constitutive equation, there is a protophysical principle known as equipresence. An independent variable present in one constitutive equation must be present in all constitutive equations.[11] In the case of Hookean solids, this condition is satisfied trivially, since the associated theory relies on a single constitutive equation.

Closing remarks

Limitations imposed by the restrictions that define continuum mechanics are also its strength. General principles together with constitutive relations constitute a theory to be studied on its own, as well as to provide a quantitative structure for such disciplines as mechanical engineering and seismology.

For both mechanical engineering and seismology, constitutive relations provide an appealing immediacy of concepts, such as rigidity of a solid. However, this immediacy is—by itself—insufficient to justify a concept within a theory.

For instance, the concept that nature abhors a vacuum, mentioned in Section 2.1.3.1, could be restated in a nondramatic vocabulary as, say, nature avoids a vacuum. It could be proposed as a simple explanation for everyday phenomena, similarly to a solid being a material that resists a change of shape. Both concepts share their empirical origins, since neither of them is deduced from an underlying theory. However, the former concept is inconsistent with the known causes, and hence, is not an acceptable epistemological tool. Blaise Pascal already knew that it was the weight of the air that caused the absence of a vacuum in our daily experience. The

---

[11]Readers interested in the equipresence principle might refer to Rivlin (1996).

Fig. 4.1

concept of rigidity, however, is consistent with the strength of bonds at the atomic scale.

Beyond hypothetical constitutive relations, the abhorrence concept becomes an ontological issue if we use it in reference to the essence of the physical realm. According to Democritus, vacuum—together with atoms, which move therein—is an essential component of reality. However, Aristotle denied the existence of an empty space. His views prevailed until the beginning of modern times and led to the ontological belief that nature abhors a vacuum, expressed with this emphatic verb. Some physicists still argue that there is no vacuum since there is a background radiation permeating spacetime.

Be that as it may, the fact that nature abhors a vacuum is supported by examining the behaviour of Earl, who is the cat in Figure 4.1. The phrase, however, loses its meaning in translation into other languages, and hence—being expression-dependent—does not qualify as a general law.

Chapter 5

Hookean solids

Science consists of facts and theories. Facts and theories are born in different ways and are judged by different standards. Facts are supposed to be true or false. They are discovered by observers or experimenters. [...] Theories have an entirely different status. They are free creations of the human mind, intended to describe our understanding of nature. Since our understanding is incomplete, theories are provisional. Theories are tools of understanding, and a tool does not need to be precisely true to be useful. Theories are supposed to be more-or-less true, with plenty of room for disagreement.

Freeman Dyson (2014)

Preliminary remarks

Having discussed, in Chapter 3, primitive concepts that underlie continuum mechanics, and, in Chapter 4, the structure of this two-part theory, we proceed with an examination of a specific theory used to study seismic phenomena, which is the theory of Hookean solids.

Hookean solids are hyperelastic materials, which are also referred to as Green-elastic materials in honour of George Green.[1] They are distinguished from other continua by the property that their stress-strain relation is derived from the strain-energy function,[2] as a consequence of ideal elasticity. Hyperelastic materials are a special case of the simple elastic materials, which are also referred to as Cauchy-elastic materials. Their stress-strain

---

[1] Readers interested in the times, work and life of George Green might refer to Cannell (2001).

[2] *see also*: Slawinski (2015, Section 4.1)

relation cannot, in general, be derived from the strain-energy function.[3] The distinguishing property of simple elasticity is less restrictive than of hyperelasticity, since—for simple elastic materials—it is required that the state of stress at any instant be determined solely by the deformation at that instant. In other words, the behaviour of simple elastic materials is not affected by the history of their deformation.

We begin this chapter by introducing the defining quantities of a Hookean solid: its elasticity parameters and its mass density. Subsequently, we examine the relationships among parameters imposed by a Hookean solid expressed as an elasticity tensor. We conclude this chapter with a discussion of models imposed by choices of material symmetries and, hence, the limits of applicability.

5.1   Models of physical realm

Constitutive relations, such as Hooke's law, allow us to connect the general theory to empirical experience within the physical realm. As discussed in Section 4.2, general principles—as their name suggests—apply to all continua. Consequently, these principles cannot introduce distinctions among continua. As stated by Epstein (2010), they are

> [...] the universal laws of Continuum Mechanics, that is, those laws that apply to all material bodies, regardless of their physical constitution. They are equally valid for solids, liquids and gases of any kind.

The property that distinguishes solids and liquids, which is their resistance, or its lack, to the change in shape, is observable and is expressed by distinct constitutive relations.

As mentioned in Section 4.1, the necessity for existence of constitutive relations, which define, and belong to, particular theories, such as fluid mechanics and solid mechanics, is a consequence of the assumption of a continuum. More specifically, it is a consequence of the fact that physical matter is not continuous in the sense proposed by continuum mechanics. Again, as stated by Epstein (2010),

> Any science must stop at some level of description, and Continuum Mechanics stops at what might be regarded as a rather high level, whereby its description of material response loses

---

[3] *see also*: Slawinski (2016, Section 3.2.3)

the character of what the philosopher of science Mario Bunge would call an interpretive explanation. This is hardly a deficiency: It is simply a definition of the point of view already conveyed by the very name of the discipline. It is this particular point of view, namely, the consideration of matter as if it were a continuum, that has paved the way for the great successes of Continuum Mechanics, both theoretical and practical. Most of Engineering is still based on its tenets and it is likely to remain so for as long as it strives to be guided by intellectually comprehensible models rather than by computer brute force. The price to pay is that the representation of the material response is not universal but must be tailored to each material or class of materials. This tailoring is far from arbitrary. It must respect certain principles, the formulation of which is the job of the constitutive theory.

These principles are discussed in Sections 4.3.2–4.3.5, above.

Constitutive relations are equations whose parameters provide a model of a physical material. For a Hookean solid, its mechanical properties are specified by elasticity parameters.

5.2 Mathematical expressions

Robert Hooke, while working on springs,[4] observed a linear relation between forces and deformations. Not surprisingly—for small deformations—a first-order approximation results in an empirically adequate mathematical model.

As illustrated in Figure 5.1, Hooke's law can be modelled by a proportionality between the extension, x, of the spring and the magnitude of force, F, applied to it by the suspended weight. The proportionality constant, k, is the elasticity parameter, which is commonly called the spring constant and whose value describes the stiffness of the spring. As discussed in Section 3.4.2, the spring is a physical model of a Hookean solid. Quantitatively, this solid is given by k, together with a mass density; both these values are real numbers, which belong to the analytic continuum, discussed in Section 3.1.

---

[4]Readers interested in Hooke's experiments, life and times might refer to Chapman (2005).

Fig. 5.1

A mathematical model, as discussed in Section 3.4.4, of the one-dimensional case of Hooke's law is

$$F = kx, \qquad k > 0, \tag{5.1}$$

which we can interpret as follows. The larger the value of k, the more force is needed to extend the spring by x; F corresponds to the force required to produce that extension. Since k is assumed to be a constant, the model is linear.

Most seismic phenomena, however, occur in three-dimensional media. To consider the stiffness of a three-dimensional Hookean solid, one needs parameters that describe the solid's resistance to changes in shape and volume due to stresses acting in different directions. Such a solid is described by its elasticity tensor.

It is a fourth-rank tensor, which means that, in three spatial dimensions, it has $3^4 = 81$ components. However, as a result of several assumptions[5]—such as that all work used in deforming a material is stored within it and ready to restore that material to its original shape—the elasticity

---

[5] *see also*: Bóna and Slawinski (2015, Appendix B.2) and Slawinski (2015, Chapters 3 and 4)

tensor, $c_{ijk\ell}$, is defined by twenty-one independent parameters that relate two-second rank tensors: the stress tensor, σ, and the strain tensor, ε.

Mathematically, the three-dimensional case of Hooke's law is expressed as

$$\sigma_{ij} = \sum_{k,\ell=1}^{3} c_{ijk\ell}\varepsilon_{k\ell}, \qquad i,j = \{1,2,3\}. \tag{5.2}$$

Herein, the second-rank tensors are symmetric, $\sigma_{ij} = \sigma_{ji}$ and $\varepsilon_{k\ell} = \varepsilon_{\ell k}$; hence, each has six independent components, not $3^2 = 9$, as one would expect, in general, for that rank. These are consequences of assumptions, respectively, the strong form of Newton's third law[6] and infinitesimal deformations.

In view of the definition of σ, together with the definition of the stress vector, $\mathbf{T}$, introduced in Section 3.2, the stress tensor has units of force per unit area: N/m^2—newtons per meter squared. The strain tensor is unitless. Hence, the elasticity tensor has units of N/m^2.[7] Commonly, this tensor is referred to as the stiffness tensor, since the larger the value of a component, the stiffer the material for a particular direction-dependent deformation.

One might expect that, in a single dimension, expression (5.2) becomes expression (5.1). Such a reduction, however, must be formulated carefully, since F is not equivalent to σ, and x is not equivalent to ε. Consider the units of expression (5.1). Unlike the unitless strain tensor, x has units of length, which entails that k has units of N/m. In particular, the stress tensor is a force per unit area; hence, its definition becomes singular in one dimension.

Thus, only a partial meaning of expression (5.2) is contained in expression (5.1). There remain important distinctions. For seismologists, to study propagation of waves within materials, expression (5.1) is not an adequate mathematical model.

Together with the corresponding equation of motion, $F = m\,\mathrm{d}^2x/\mathrm{d}t^2$, expression (5.1) results in an ordinary differential equation,

$$\frac{\mathrm{d}^2x}{\mathrm{d}t^2} = -\frac{k}{m}x, \tag{5.3}$$

which is equation (3.4). Herein, the negative sign is introduced by interpreting Hooke's law in terms of the force required to regain the original length of the spring; it is the force that generates the oscillatory motion. The solution of this equation is a motion—as a function of time alone—of

---

[6] *see also*: Slawinski (2015, Section 2.7)
[7] *see also*: Slawinski (2015, Section 3.2.2)

the mass attached to the end of the spring, as illustrated in Figure 5.1; it is not enough information to study wave propagation.

On the other hand, together with the corresponding equation of motion, stated in expression (6.1) on page 105, expression (5.2) results in a partial differential equation that, in a single spatial dimension, becomes

$$\frac{\partial^2 u(x,t)}{\partial t^2} = \frac{k}{\rho}\frac{\partial^2 u(x,t)}{\partial x^2}, \tag{5.4}$$

which is the wave equation, whose solution is a displacement as a function of both time and position.

The physical model for equation (5.4) can be an elastic string, such as for the cello illustrated in Figure 6.1 on page 107. Equation (5.4) refers to every point of such a string; equation (5.3) refers only to the position of the mass at the end of the spring.

Herein lies the essence of continuum mechanics and of wave propagation within mechanical continua. Equation (5.4) refers to deformation of the string; equation (5.3) refers to motion of mass m. In the latter case, no deformation is examined. To consider the entire spring, the deformation must be given for each point as a function of time, $u(x,t)$, and the mass as its density at each point. In contrast to equation (5.3)—in which mass refers only to the object at the end of the spring, and the spring itself is assumed to be massless—the mass of the string is distributed through that string.

It is possible to achieve a continuum-mechanics model in the context of expression (5.1), by replacing spring by a string, by letting mass be the product of length of the string and its mass density, and by including the deformation as $\partial u(x,t)/\partial x$, instead of x.[8] However, for Hooke's law in the single spatial dimension, there remains the issue of the aforementioned singularity for a stress tensor to replace F on the left-hand side of expression (5.1). This issue can be formally avoided by proceeding to the equations of motion using the right-hand side of expression (5.1).[9]

In general, as a result of these requirements, the mathematical models of continuum mechanics are necessarily expressed by partial differential

---

[8] *see also*: Slawinski (2016, Section 5.1.3)

[9] Readers interested in wave propagation in an elastic bar might refer to Roberts (1994, Section 4.2.1), prior to which he writes the following aside.

> Note that Hooke's law was originally intended to apply only to the stretching of a rod as a whole. However, we are going to use the law as if it applies to every small segment of the rod individually.

equations, such as equation (5.4), and in terms of tensors, such as expression (5.2). As emphasized by Epstein (2010), the discourse of continuum mechanics is in the language of differential geometry.

Remark 5.1. Tensors are a generalization of scalars, which are described by their magnitude, and vectors, which are described by both magnitude and direction. In view of Cauchy's stress principle stated on page 63, the stress tensor, which is a second-rank tensor, relates the stress vector to the orientation of a surface on which it is acting; thus, there is both direction and orientation to consider. Within this generalization, scalars are tensors of rank zero, and vectors are tensors of rank one. Hence, the stress tensor relates linearly two first-rank tensors. In an analogous manner, the elasticity tensor, which is of rank four, relates linearly two second-rank tensors.

Hookean solids—similarly to the fictional character of Sherlock Holmes, discussed on page 55— possess mass. This mass is required by the primitive concept of a material body, as discussed in Section 3.2.3. In accordance with expression (3.2), the mass within a volume is obtained by specifying the density, ρ.

For a given Hookean solid, it is common to include the mass-density information by dividing the elasticity parameters by the value of ρ. Such a division results in units of these parameters being m^2/s^2.

Let us revisit the issue of simplicity discussed in Section 1.3.2. As a linear relation between the stress and strain tensors, expression (5.2) can be viewed as the simplest case.[10] However, even though expression (5.2) is algebraic and linear, the resulting expressions for the travel time of wave propagation are nonlinear, and involve transcendental functions, such as logarithms, which cannot be expressed in terms of a finite sequence of the algebraic operations of addition, multiplication and root extraction.[11] A more complicated constitutive relation might lead to simpler travel-time expressions used by seismologists. In other words, one could strive to postulate a relation whose form entails a simplicity of subsequent formulæ.

---

[10]Readers interested in nonlinear elasticity for quasistatic processes might refer to Ogden (1997/1984), where the author presents

> the mathematical theory of non-linear elasticity [...] and the analysis of the mechanical properties of solid material capable of large elastic deformations.

Seismologists, however, are mostly concerned with dynamic processes due to wave propagation, and with rocks, which are not capable of large elastic deformations.

[11]*see also*: Slawinski (2015, Section 14.3.3)

The wealth of mathematical physics contained within and examined by the theory of only linear elasticity is impressive (Achenbach, 1973; Marsden and Hughes, 1983). This wealth justifies the fact that seismologists base a large portion of their work on that theory (Aki and Richards, 2002).

5.3   Symmetries

5.3.1   *Symmetry types*

Since symmetries of Hookean solids are mathematical properties, let us comment on the status of symmetries in the context of applied mathematics, as discussed in Section 3.3, above. A rotational symmetry of a crystal and a translation symmetry of a geological layer are properties of the physical realm, even though they might be modelled by group theory, which resides in the mathematical realm.[12] By idealizing the Earth as a continuum, seismologists introduce symmetries that do not explicitly exist in the Earth. They are symmetries of the elasticity tensor.[13] These symmetries are operations that leave the tensor unchanged. Let us consider two types of symmetry of a Hookean solid.

5.3.2   *Index symmetry*

[14]For Hookean solids,

$$c_{ijk\ell} = c_{jik\ell} = c_{k\ell ij} \,, \tag{5.5}$$

which are index symmetries that reduce the number of independent components from eighty-one to twenty-one. Herein, the symmetry means that we can exchange i with j and k with ℓ, as well as the pairs ij with $k\ell$, without modifying the tensor.

Mathematically, the first and second symmetry are the consequence of the symmetry of stress and strain tensors, respectively. The third one follows from the equality of mixed partial derivatives.

These mathematical properties are expressions of physical assumptions. The first one is a result of assuming the strong form of Newton's third law[15]

---

[12]Readers interested in relations between group theory and the physical realm might refer to Zee (2016).
[13]*see also*: Slawinski (2015, Chapter 5)
[14]*see also*: Slawinski (2015, Sections 3.2.2 and 4.2.1) and Slawinski (2016, Section 3.2.4.3)
[15]*see also*: Slawinski (2015, Section 2.7)

or of taking the balance of linear and rotational momenta as postulates, and the second one is a result of assuming infinitesimal—as opposed to finite[16]—deformations only. The third symmetry is a consequence of assuming that all energy expended on deformation is stored—as a potential energy—in the deformed material.[17]

According to expression (5.5), $c_{1112} = c_{1121} = c_{1211} = c_{2111}$, and so on. Such relationships reduce the eighty-one components of a fourth-rank tensor to twenty-one independent parameters of an elasticity tensor. This reduction is a necessary condition for a fourth-rank tensor to represent a Hookean solid. Not every fourth-rank tensor represents a Hookean solid, even though every Hookean solid is represented by a fourth-rank tensor. The other condition is that the tensor be positive-definite, which is a mathematical statement of conservation of energy, referred to, in the context of continuum mechanics, as the stability condition, since it implies that a deformable solid remains in its original shape, unless work is performed to deform it.[18]

5.3.3 *Material symmetry*

5.3.3.1 *Symmetry classes*

[19]The values of the components of the elasticity tensor are direction-dependent, which means that, in general, Hookean solids are anisotropic, even though a particular Hookean solid can be isotropic.

Considering group theory to examine material symmetries, the group of a given symmetry class contains the rotations of the coordinate system that leave the tensor invariant.[20] For an isotropic Hookean solid, the symmetry group contains all rotations. Expressed in terms of coordinates, such an invariance reduces the number of linearly independent elasticity parameters, beyond the twenty-one parameters discussed in Section 5.3.2. For instance, for an isotropic Hookean solid, $c_{1111} = c_{2222} = c_{3333}$, $c_{1212} = c_{1313} = c_{2323}$, and so on; consequently, there are only two independent parameters.

Between general anisotropy and isotropy, there are six symmetries of the Hookean solid: monoclinic, orthotropic, tetragonal, transversely isotropic, trigonal and cubic (Bóna *et al.*, 2004b). Herein, the symmetry means that

---

[16] *see also* : Slawinski (2016, Section 3.2.3)
[17] *see also* : Slawinski (2015, Section 4.2)
[18] *see also* : Slawinski (2015, Section 4.3)
[19] *see also* : Slawinski (2015, Chapter 5) and Slawinski (2016, Chapter 3)
[20] *see also* : Slawinski (2015, Section 3.2.4.1)

we can rotate the coordinate system without changing the values of the components of the elasticity tensor, which—in turn—implies that the mechanical properties remain unchanged under these rotations.

For instance, the monoclinic symmetry means that there is one symmetry plane. We can exchange the positive and negative directions of one axis without affecting the values of the components of the tensor and the mechanical properties of the Hookean solid it represents. Tetragonal symmetry exhibits components that are invariant under rotations by $\pi/2$, which means that the solid has a fourfold symmetry about an axis.

[21]It follows from a theorem of Herman (1945) that no Hookean solid can have a fivefold, sixfold, sevenfold, or higher, symmetry about any axis (Bóna *et al.*, 2004b). This theorem states that the largest-fold symmetry that any tensor can exhibit is given by its rank. Thus, for the elasticity tensor, the fourfold symmetry about a given axis is the largest discrete symmetry that it can exhibit. Beyond the fourfold symmetry, there is invariance under any rotation about that axis. Hookean solids that exhibit such a property are called transversely isotropic.

The existence of continuous symmetry groups of Hookean solids—namely, isotropy and transverse isotropy—is a result of mathematical modelling in terms of continuous objects expressed by tensors. This is one of the properties that distinguishes symmetries of Hookean solids from crystal symmetries. No continuous symmetries appear in crystals, whose symmetries refer to lattices, which are discrete objects.

The six classes that lie between general anisotropy and isotropy exhibit a partial ordering (Bóna *et al.*, 2004a).[22] It is not a progressive increase of symmetries. A tetragonal solid exhibits more symmetries than an orthotropic one and fewer than a transversely isotropic one, but it has different symmetries than a trigonal one. In other words, one cannot add symmetry-group elements to the trigonal case to make it tetragonal or vice-versa.

Thanks to this partial ordering, we do not lose the distinction between the cases for which symmetry planes are or are not orthogonal to each other. For the tetragonal case, we have symmetry planes orthogonal to each other and for trigonal they intersect at $2\pi/3$.

---

[21]*see also*: Slawinski (2015, Section 5.10.2 and Exercise 5.8) and Slawinski (2016, Section 3.2.4.14)

[22]*see also*: Slawinski (2015, Figure 5.13.1) or Slawinski (2016, Figure 3.3)

5.3.3.2 *Limitations of material symmetry*

Many patterns of actual materials have no analogies in symmetries of Hookean solids. Even though, for an elasticity tensor, all symmetries greater than fourfold about a given axis appear as transverse isotropy, this is not the case for discrete symmetries, such as symmetry of crystals, whose classification allows for hexagonal and octagonal symmetries. In the context of Hookean solids, one cannot distinguish between, say, hexagonal and octagonal symmetries; it would require a different elasticity theory, whose constitutive relation contains a tensor of rank eight.[23] The tradeoff between accuracy and idealization is a recurring theme.

Thus, using Hookean solids to model a physical material as transversely isotropic does not imply that this material has the same behaviour in all directions about the rotation axis, since the material and the model are not isomorphic. Also, the impossibility of a Hookean solid having only two orthogonal symmetry planes is a consequence of the even rank of the elasticity tensor; it does not imply that a material exhibiting such symmetries cannot exist in the physical world. In both cases, the analogy is subject to the limitation of the theory used to model the material.[24]

As discussed in Section 3.3, there is a homomorphism between the mathematical realm and the continuum represented by Hookean solids. Hence, a given property of the elasticity tensor is a property of the corresponding Hookean solid. However, the Earth need not exhibit that property, since the Earth is not isomorphic to the Hookean solid.

Remark 5.2. An isomorphism is a special type of homomorphism. It is a homomorphism that is both one-to-one and onto. In other words, an isomorphism is a bijective homomorphism.

If there is an isomorphism between two entities, then their structures correspond completely to one another. If there is a homomorphism between two entities, then only parts of their structures correspond to one another. Thus, for the former, no information is lost. For the latter—which is a more common relation between the physical and abstract realms—there is a loss of information.

Neither mathematics nor seismology can impose a unique model. There is a range of suitable options. The choice of a tensor rank illustrates the freedom involved in the choice of a mathematical structure to represent a physical phenomenon.

---

[23] Readers interested in further discussions of these issues might refer to Slawinski (2014).
[24] *see also*: Slawinski (2016, Sections 3.2.4.13 and 3.2.4.14)

There is another pragmatic limitation of an epistemological nature. One must be careful drawing conclusions about the three-dimensional physical world based on considerations of fewer physical dimensions, even though the latter might be easier to visualize or use in calculations.

Material symmetry classes, including their number, depend on the spatial dimension. A Hookean solid in a single spatial dimension is isotropic, since—in view of the elasticity tensor being of even rank—the properties are invariant to the sense along a given direction. In two spatial dimension, there is no distinction between isotropy and transverse isotropy, which appears in three dimensions. A trigonal symmetry cannot exist in two dimensions, since it would violate the so-called point symmetry, which is the invariance under all reflections about the origin, and is another consequence of the even rank of the elasticity tensor.

5.3.3.3   *Usefulness of material symmetry*

Having acknowledged certain limitations of material symmetries of Hookean solids, in Section 5.3.3.2, let us comment on their usefulness. If the aim of science is merely empirical adequacy, then there is nothing to be gained by information about symmetry, since a generally anisotropic model described by twenty-one parameters is necessarily more accurate than a symmetric model with fewer parameters, in particular, much more accurate than an isotropic model, which is described with only two parameters. However, if we consider the aim of seismology to be the description of the Earth, then the information about symmetries gives us information about the material represented by a Hookean solid. For instance, a transversely isotropic model might suggest a series of parallel layers (Backus, 1962). An isotropic Hookean solid might indicate a material exhibiting a random structure, since symmetries are associated with patterns.[25] Thus, by considering symmetries, we infer information about Earth's materials.

To gain further insight into searching for symmetries rather than solely accounting for measurements by remaining at the level of general anisotropy, consider the fact that it might be preferable to represent a cloud of points as a straight line, rather than to account for every point by invoking a high-order polynomial. By considering symmetries, seismologists might recognize geological trends that otherwise would be lost in details.

---

[25] Readers interested in physical interpretation of symmetry classes might refer to Slawinski (2014).

Furthermore, accounting for all details is not tantamount to accuracy, and seismologists strive to construct a model to describe a physical material, not a model to describe the acquired data set.

Another justification to consider a symmetric Hookean solid is the resulting simplicity of description. Fewer elasticity parameters result in simpler equations that seismologists need to consider. There is no need to increase the complexity, if an adequate description can be obtained using a simpler model.

5.4 Model limitations

5.4.1 *Thermal effects*

[26]In postulating Hookean solids, seismologists constrain their studies to isothermal processes, even though propagation of mechanical disturbances with the Earth is neither isothermal nor adiabatic. Such propagation results in a change of temperature of a portion of the material and there is a heat transfer between a given portion of the material and its surroundings. Theoretical processes modelled within Hookean solids are not explicit functions of temperature changes or heat transfer.

To evaluate such a simplifying assumption, it is necessary to compare the speed of propagation of deformations and of thermal diffusion. As it turns out, most seismic events are approximately adiabatic. The heat exchange between the deformed material and its surroundings is negligible during wave propagation due to both the shortness of duration of the process and the low thermal conductivity of most rocks. By contrast, the postglacial rebound is approximately isothermal, since there is a thermal equilibrium between the deformed material and its surroundings.

However, since the difference in values of the experimentally determined elasticity parameters using the isothermal and adiabatic approaches is negligible, seismologists can use either one.[27] The choice is dictated by the convenience of explicitly temperature-independent parameters.

This explicit independence does not mean that—for a given material— the elasticity parameters are the same for all temperatures. The rigidity of granite is different at different temperatures, say, 0° and 500°. This difference in rigidity is expressed by different values of parameters of two Hookean solids used to study the behaviour of granite in two different

---

[26]*see also*: Slawinski (2016, Section 8.3)
[27]*see also*: Slawinski (2015, Section 8.3.2)

settings. However, it does mean that—for a given ambient temperature—
the heating effect of wave propagation is ignored, since it is considered
negligible.

One can consciously adopt a false model if one is aware of how and why it
departs from reality. Herein, aware of negligible discrepancy, seismologists
choose the convenience of the isothermal process to obtain the values of elasticity parameters, even though seismic events are approximately adiabatic.

5.4.2   *Interaction between waves and continua*

Beside the neglect of thermal effects of wave propagation, discussed in Section 5.4.1, there are other idealizations and limitations within seismology
that result from postulating Hookean solids. Let us discuss another one of
them.

Mass density and elasticity parameters of a Hookean solid determine
properties of wave propagation within the medium, such as velocities of
propagation as well as amplitudes and directions of displacements. On the
other hand, within the context of a Hookean solid, properties of the medium
are not affected by its deformations due to wave propagation. This is a
consequence of describing any point within a continuum by constant values
of the elasticity parameters. To include the effect of wave propagation on
material properties, it would be necessary to modify expression (5.2) by
including the effect of deformation. This could be achieved by letting $c_{ijk\ell}$
be a function of $\varepsilon_{k\ell}$. However, this—at first sight minor—modification
would have several consequences.

First, one would have to find an appropriate form of such a function.
For instance, one might distinguish the interaction between the continuum
and the wavefront of the primary wave from the interaction between the
continuum and secondary waves, since in the latter case the continuum is
already affected by the passage of the wavefront.

Also, the modified expression would be nonlinear.[28] Thus, we accept
that, even though certain aspects of physical phenomena are ignored, expression (5.2) is empirically adequate.

---

[28]Readers interested in nonlinear elasticity might refer to Ogden (1997/1984), where the
author presents

> [t]he incremental theory governing the change in deformation due to
> a small change in material properties,

which can be expressed in terms of hysteresis effects, ignored in the theory of seismic
waves.

Empirical adequacy is a common criterion. For instance, for a pendulum, equation (3.4), on page 72, is linearized under the assumption of small oscillations, by letting $A\sin x = Ax$, which is an approximation for small angles expressed in terms of radians. Without this linearization—and, hence, without limiting the examination to small angles—there is no solution of equation (3.4) in terms of elementary functions.

There is an interesting analogy between the interaction of waves with the medium and Newton's absolute space. According to Newton's first law of motion, space allows bodies to move at constant speeds in straight lines—provided no other force is acting upon them—but bodies have no effect on space. By contrast, according to general relativity, space and matter interact with one another.

Seismology, by allowing the material to affect waves without waves affecting the material, would seem to disagree with the accepted view of physics. However, in seismology, this one-way causation is an idealization, as is the concept of continuum itself. Newton's absolute space, by contrast, is not an idealization but was intended to be a direct description of reality.

Closing remarks

In mediating a physical interpretation of seismic data by seismological theory, Hookean solids play a role of analogies. They are analogies, not approximations, since a rock cannot—even in the limit—approach a Hookean solid. A Hookean solid and a rock belong to different realms: abstract and physical, respectively.

Commonly, the adjustment of such an abstract model to agree with seismic data is achieved by adjusting the parameters of Hookean solids. According to Popper, such adjustments might become an issue for his refutability criterion of science, as discussed in Section 1.4.2. According to Kuhn, they are part of the normal science, discussed in Section 2.1.4.2.

Chapter 6

Forward and inverse problems

La tradition veut que deux branches scientifiques soient considérées, la branche théorique et la branche expérimentale. [...] De cette approche systémique naît le concept de modélisation. [...] Ces trois piliers de la science, théorie, expérience et simulation, sont maintenant indissociables et chaque avancée dans l'un ou l'autre domaine se répercute rapidement sur les deux autres.[1]

Matthieu Lefebvre (2012)

Preliminary remarks

To study seismological problems, seismologists need to consider propagation of deformations that results in recording of the disturbances emanating from a distant earthquake. Given the location of the source of disturbances and assuming the properties of Hookean solids between that location and a sensor, seismologists can model the expected behaviour recorded at the receiver. Alternatively, examining information at the sensor, seismologists strive to infer properties of the medium separating the source from the receiver. These approaches are referred to as forward and inverse problems, respectively.

We begin this chapter by considering propagation of disturbances within a Hookean solid. Subsequently, to deal with a forward problem, we discuss its constraints provided by the initial conditions, which entail observables.

---

[1]Traditionally, two branches of science are considered: theoretical and experimental. [...] From this systemic approach arises the concept of modelling. [...] These three pillars of science, theory, experiment and modelling, are presently inseparable, and an advancement in one affects promptly the other two.

We conclude this chapter with a discussion of inverse problems and their intrinsic limitations, such as lack of uniqueness of solution.

6.1 Inference of terrestrial properties

Once a theory is established, say, Newton's gravitation or seismic ray theory, we can make predictions concerning motion of planets or the time it takes a disturbance to propagate from the earthquake location to a distant seismograph. Within differential equations that describe the wave motion, the process is deterministic. Hence, the information about planets or seismic disturbances is obtained by deductive argumentation.

Determining the Earth's structure is an inverse problem, which, in general, is more challenging than its forward counterpart. Given the locations of the source and receiver together with information about mechanical properties of the medium, the theory of continuum mechanics allows us to predict uniquely the travel time and displacement of the seismic signal recorded at the receiver. However, given the data of recorded travel time and displacement, we cannot—in general—obtain the location of the source and learn about mechanical properties of the medium. Furthermore, even if the locations of the source and receiver are known, it is impossible to deduce—in general—properties of the medium. It might be possible to do so in particular cases; for instance, if we assume that the Hookean solid is homogeneous, which is a very constraining assumption and might not be of interest for a realistic model of the Earth. Obtaining information about medium properties is an inductive process.

Yet, the information about properties of an anisotropic inhomogeneous Hookean solid is not equivalent to predicting the internal structure of the Earth. Further inference or interpretation is required. Determining the existence and size of a solid inner core, discussed in Example 1.4 on page 15, is the result of an inductive process about properties of continua, combined with interpretation about terrestrial properties.

Herein, we postulate the fourth principle of seismology.

IV. Contact between the constitutive equations and the Earth is made through seismic measurements whose interpretations are within the theory of seismology based on the quantitative framework of continuum mechanics.

This principle is the link between the theory and its various models, on the one hand, and the empirically accessible physical realm, on the other.

Additional concepts might be required to establish this link, such as the theory underlying the functioning of a detecting device. For instance, electromagnetic induction might be used to record the motion of a seismograph invented in 1906 by Boris Borisovich Golitsyn.[2] Assuming that this is understood, this principle provides the bridge between the Earth and various hypotheses of continuum mechanics.

The justification for this principle is the agreement between measurements and predictions provided by continuum mechanics. Even though issues of nonuniqueness of solutions appear, it is reasonable to proceed with a realistic interpretation of seismic data. In turn, the reasonableness of that interpretation is the *a posteriori* justification of seismology as a hypotheticodeductive scientific discipline. Other considerations are also relevant, such as consistency with other theories, say, cosmogony.

The four principles, stated on pages 74, 74, 84 and 104, complete the foundations of seismology announced on page 5. Seismology is a science concerned with the mechanical properties of the Earth and Earth-like planets. It is not concerned explicitly with—although it might prove useful to—other branches of geoscience.

6.2 Forward problem

To consider a forward problem for the propagation of seismic waves, we invoke the equation of motion,

$$\sum_{j=1}^{3} \frac{\partial \sigma_{ij}}{\partial x_j} = \rho \frac{\partial^2 u_i}{\partial t^2}, \qquad i \in \{1, 2, 3\}, \tag{6.1}$$

where x_j and t denote space and time, respectively. In the context of the two-part theory, discussed in Section 4.1, equation (6.1) belongs to the general theory, which is valid for all continua. The left-hand side of this equation is the expression of force due to stress, σ_{ij}. The right-hand side, where ρ is mass density and u_i is a component of the displacement vector, is—within an infinitesimal volume of a continuum—the expression of force as the product of mass and acceleration.

To consider propagations within Hookean solids, we invoke expression (5.2) and expresses σ_{ij}, in equation (6.1), in terms of $c_{ijk\ell}$.[3] Since the

---

[2] Readers interested in the life of Prince Golitsyn and the fate of his family following the Soviet revolution might refer to Smith (2012).

[3] *see also*: Slawinski (2015, Chapter 7)

mass density, ρ, and the elasticity parameters, $c_{ijk\ell}$, are the only quantities needed to describe a Hookean solid, the resulting equation, which is the elastodynamic equation, contains the information about a given elastic medium.

Equation (6.1) is a differential equation derived from the balance of linear momentum, which—in equation (4.2)—is stated in terms of integrals. There is an important simplification between equation (4.2) and equation (6.1). In the latter, we ignore body forces, such as gravity,[4] whose effect on seismic propagation—within the range of frequencies exhibited by seismic waves—is several orders of magnitude smaller than the effect of elasticity.[5] This is a statement of empirical adequacy of equation (6.1); it does not imply that body forces have no effect at all.

The fact that—for the equation of motion in the context of seismology—the effect of forces associated with elasticity is so much greater than the effect of gravitational forces means that the terrestrial and lunar behaviour of seismic waves depends on mechanical properties of the terrestrial and lunar materials but is not affected—in an empirically detectable manner—by different magnitudes of gravitational forces on the Earth and on the Moon.

Similarly, and as illustrated by the comic strip in Figure 6.1, in view of the negligibility of the gravity effects for higher frequencies, a cello—whose lowest frequency is $C_2 = 65.4$ Hz—tuned on the Earth would vibrate with nearly the same frequencies on the Moon or in any other place where we might take it, including the areas of weightlessness. To a disappointment of the cellist and the audience, the sound generated by these vibrations is not transmittable through the void, so the music would not be heard without an atmosphere, but the instrument would remain well-tuned.

There is another modification between equations (4.2) and (6.1); the former is an integral equation and the latter is a differential equation.[6] To derive a differential equation from an integral equation, one uses the divergence theorem, which imposes constraints on the solutions that are not required in the integral formulation, in particular, differentiability of functions involved.[7] A common use of differential equations in physics is, at least to a certain extent, of a historical nature, since theories of differential equations have been more developed than the theory of integral equations.

---

[4]Readers interested in the effect of gravity in the context of equation (6.1) might refer to Müller and Weiss (2016, Section 1.2.1).
[5]*see also*: Slawinski (2016, Section 8.3)
[6]*see also*: Slawinski (2015, Section 2.6.1)
[7]*see also*: Bóna and Slawinski (2015, Appendix A.1)

Fig. 6.1

The elastodynamic equation contains the possible propagations of disturbance in a Hookean solid in accordance with the general principles. To study a particular event, we identify the actual propagation by providing initial conditions, such as the location and the velocity of displacement at a given instant, which are commonly referred to as Cauchy data.

The information provided by Cauchy data need not correspond to the initial instant. It could describe the disturbance at any instant. All that is required is that these conditions state the solution at a single instant. Given the elasticity parameters and mass density, the elastodynamic equation allows us to describe the propagation that follows or precedes the instant for which the aforementioned information is provided.

Conditions that supplement differential equations, such as initial conditions, play a role akin to constitutive equations in continuum mechanics. They allow us to consider a particular case among all physically allowable cases. Also, neither stems from fundamentals. Both initial conditions and constitutive equations might be formulated from empirical knowledge, provided they do not contradict the general concept, whose particular case they are to elucidate.

The elastodynamic equation belongs to the realm of the initial-value problems, known as the Cauchy problem (Hadamard, 1932, pp. 3-26), (Trèves, 2006, pp. 89-186). According to the Cauchy-Kovalevskaya theorem,[8] the solution of a Cauchy problem exists, is unique, and depends continuously on the Cauchy data, which means that a small change within initial conditions results in a small change in the solution. The existence, uniqueness and stability of a solution are the requirements for a complete description of a physical problem, and the well-posedness of hyperbolic partial differential equations, in the sense of Hadamard, requires the initial conditions (Hadamard, 1932, pp. 10-11), (Trèves, 2006, pp. 142-155). In contrast to inverse problems, discussed in Section 6.3, below, the forward problems are well-posed.

Remark 6.1. Beside the absence of body forces, the form of equations of motion invoked in seismology also have an important simplification, namely, ignoring the distinction between two descriptions of motion within a continuum. Since a continuum, by its definition, does not exhibit discrete elements, the description of motion within it faces issues that are not encountered in describing the motion of a particle. Commonly, there are two descriptions: material and spatial.[9]

Heuristically, the former can be illustrated by following the motion of a droplet of dye in a flow. The latter can be illustrated by considering the flow through a fixed spatial location, such as a river passing through a dam. The one-to-one relation between these descriptions requires that distinct elements within a continuum cannot—at the same instant—occupy the same spatial location.[10]

The assumption of infinitesimal displacement entails that the two descriptions are—to the first order—equivalent to one another. Hence,

---

[8] *see also*: Bóna and Slawinski (2015, Section 1.4)
[9] *see also*: Slawinski (2015, Section 1.3) and Bóna and Slawinski (2015, Appendix B.1)
[10] *see also*: Slawinski (2015, Section 1.2.2)

seismologists—unlike fluid mechanists, who study finite displacements—can ignore this distinction.

6.3 Inverse problem

A formulation of, and a solution to, a forward problem is necessary, but not sufficient, conditions for the solubility of the inverse one. Most inverse problems are mathematically ill-posed, in the sense that the solution is not unique, since many input scenarios can result in the same output. The term ill-posed results from the antiquated belief of determinability. According to Pierre-Simon de Laplace (1812),

> L'état présent de l'univers est l'effet de son état antérieur, et la cause de ce qui va suivre. Une intelligence qui, à un instant donné, connaîtrait toutes les forces dont la nature est animée, la position respective des êtres qui la composent, si d'ailleurs elle était assez vaste pour soumettre ces données à l'analyse, elle embrasserait dans la même formule les mouvements des plus grands corps de l'univers, et ceux du plus léger atome. Rien ne serait incertain pour elle, et l'avenir comme le passé seraient présents à ses yeux.[11]

We can make several comments about this quote. Herein, de Laplace (1812) places too much trust in the capacity of calculation, since—given perfect knowledge of initial conditions—there are deterministic situations within classical mechanics that cannot be calculated (Pour-El and Richards, 1983). Herein, by calculation, we mean algorithmic and computable by a computer in a finite time. One might obtain an answer in a different manner, such as a guess. This is akin to the nonsolubility of polynomials of degree greater than four by radicals, which does not imply that they have no solutions.

In considering forward and inverse problems, we do not mean, as Laplace did, predicting the future or retrodicting the past. The forward problem consists of obtaining observables, given mechanical properties, and the inverse problem of inferring mechanical properties from observables. The

[11]The present state of the universe is the effect of its past and the cause of its future. A mind that, at a certain moment, would know all forces that set nature in motion, and all positions of all items that compose her, if this intellect were also vast enough to submit these data to analysis, it would embrace in a single formula the movements of the greatest bodies of the universe and those of the tiniest atom. Nothing would be uncertain for such an intellect, the future just like the past would be present before its eyes.

ill-posedness of seismological inverse problems is an instance of the under-determination of theory by evidence, discussed in Section 1.3.1.

The issue of underdetermination can also appear in verbal communication, as illustrated by the following example.

Example 6.1. ut tensio sic vis
Consider the following problem proposed by Robert Hooke.

> The true theory of elasticity or springiness, and a particular explication thereof in several subjects in which it is to be found: And the way of computing the velocity of bodies moved by them. *ceiiinossssttuv*

Herein, *ceiiinossssttuv* is Hooke's anagram, designed to lay claim to his discovery, which he did not wish to share until he accomplished further aspects of his research. He revealed its meaning in the next passage.

> About two years since I printed this theory in an anagram at the end of my book of the descriptions of helioscopes, viz. *ceiiinossssttuv*, that is *ut tensio sic vis*; that is, the power of any spring is in the same proportion with the tension thereof: that is, if one power stretch or bend it one space, two will bend it two, and three will bend it three, and so forward.

As can be verified, *ut tensio sic vis* is composed from *ceiiinossssttuv*. There are, however, other latin phrases that can be composed with the same letters. The solution is not unique.

The inverse problem has a unique solution if and only if there is an isomorphism between observables and model parameters. Let us consider the following example to illustrate the importance of such a one-to-one and onto relation between observables and model parameters

Example 6.2. Can one hear the shape of a drum?
Let us consider a classic inverse-problem paper of Mark Kac (1966), entitled *Can one hear the shape of a drum?*, in which the reader is asked to investigate whether or not differently shaped drums can produce the same sound. The model is introduced by the wave equation in two spatial dimensions, and the reader is asked to infer from the eigenfrequencies the shape of the membrane that forms the drum.

As it turns out, the same eigenfrequencies exist for different shapes. Hence, differently shaped membranes can produce the same sound. This was not known at the time of publication of that paper. Thus, there is no unique solution and, hence, we cannot infer uniquely the shape of a drum. However, we might be able to restrict the possibilities to only a few shapes that correspond to given eigenfrequencies.

An ability of such a restriction is an important quality of inverse-problem solutions. To comment on uniqueness and restriction, let us consider the following example.

Example 6.3. My dear Watson
"My dear Watson, when you have eliminated the impossible, whatever remains, however improbable, must be the truth."

Holmes implicitly assumes that the remaining solution, however improbable, must be unique, which is exceedingly optimistic.

Geophysical inverse problems have been tackled by many mathematicians. In 1826, Niels Abel considered the problem of determining the shape of a hill, given the two-way travel time as a function of the initial velocity of a particle projected up that hill. The particle is assumed to be in a constant gravitational field.[12] A century later, Gustav Herglotz and Emil Wiechert considered obtaining the speed of a seismic wave as a function of depth given the two-way travel time of a signal as a function of position along the surface.[13] Recently, Pestov and Uhlmann (2005) solved the problem of obtaining the wave speed as a function of position within a body, given the signal travel times along the surface. The body is similar to the one illustrated in Figure 1.4, on page 12, but without assumption of circular symmetry.

These three problems are formulated and solved in two spatial dimensions. However, it appears that Stefanov *et al.* (2017) have extended the solution of the last one to three spatial dimensions.

Closing remarks

As a part of the natural sciences, whose purpose is to infer properties of nature, seismology is concerned with inverse problems. Hence, even though the forward problem, which is discussed in Section 6.2, is endowed with a deductive formulation, its mathematical certainty is not inherited by subsequent physical conclusions. These conclusions are subject to the reliability

---

[12]Readers interested in Abel's problem might refer to Aki and Richards (2002, Box 9.3).
[13]Readers interested in the Herglotz-Wiechert formula might refer to Aki and Richards (2002, Section 9.4.1).

of initial assumptions and inferred interpretations. Inverse problems in seismology, which are discussed in Section 6.3, similarly to most scientific arguments, are necessarily inductive.

The unavoidable loss of certainty of any seismological inference—beyond the statements within the realm of continuum mechanics—is stated by the inference problem of David Hume (1739/2007).[14] We cannot prove the validity of induction without *a priori* assuming the uniformity of nature, which is an example of the logical fallacy of *petitio principii*, referred to as begging the question: we presuppose properties of nature in order to prove them. Thus, the perception of causality is based on experience, which is based on the assumption that the future resembles the past, which, in turn, is based on experience, thus leading to a circular argument. In a strict sense, any causality—beyond the entailments within the mathematical realm— might be only a more or less random correlation, as illustrated by the comic strip in Figure 6.2.

One could argue that the logical fallacy of *petitio principii* can be avoided by accepting—as an axiom—the principle of uniformitarianism, mentioned in Section 2.1.6. However, such an avoidance is only formal; by postulating that axiom, Hutton and Lyell acknowledge the essence of the inference problem.

According to the principle of uniformitarianism, which underlies most geological inferences, the Earth's processes of the past acted in the same manner and, essentially, with the same intensity, as they do now. Also, approaches based on that principle extend beyond the terrestrial domain of geology. For instance, we conjecture that our Solar system was formed following an explosion of a massive star, which is referred to as a supernova, and is composed—upon processes of accretion—from the remaining dust, which is referred to as the nebula. Accepting the uniformity of nature, we interpret the nebula of the Orion constellation, which is visible in the night sky, as an earlier stage of the formation of such a system.

Foundational problems often remain unsolved or even insoluble, as might be the case of Hume's inference problem. Yet, even though this might not prevent us—in a strict sense—from drawing conclusions from seismological studies, it should make us alert to the limited reliability of inferences we make.

---

[14]*see also*: Slawinski (2016, Section 7.2.7)

Fig. 6.2

Chapter 7

Intertheory and intratheory relations

The relationship between mathematics and physics has changed profoundly over the centuries. Up to the eighteenth century there was no sharp distinction drawn between mathematics and physics, and many famous mathematicians could also be regarded as physicists, at least some of the time. During the nineteenth century and the beginning of the twentieth century this situation gradually changed, until by the middle of the twentieth century the two disciplines were very separate. And then, toward the end of the twentieth century, mathematicians started to find that ideas that had been discovered by physicists had huge mathematical significance.

Timothy Gowers (2008)

Preliminary remarks

There are many structural and conceptual relations within a given theory and among different theories. Such relations can be insightful in examining a variety of issues, and understanding them allows us to draw fruitful conclusions about these issues and helps us avoid unjustifiable expectations from a scientific formulation.

We begin this chapter by discussing the concepts of reduction and emergence. We proceed with a discussion on approximations and generalizations, and comment on the partial ordering of models. Subsequently, we compare the scientific approaches of geology and geophysics. We conclude the chapter by commenting on seismology as a classical field theory to show that the continuum invoked in seismology belongs to that category of fields, with certain ontological and epistemological qualifiers.

7.1   Reduction and emergence

There are many ways to view relations among theories. It is common to see these relations as a hierarchy based on reduction. According to this account, physics is supreme in the natural sciences, since chemistry can be reduced to physics, and biology to chemistry. Steven Weinberg (2001) is one of the prominent champions of this view.

On the one hand, the quantum mechanical explanation of the periodic table supports the reduction of chemistry to physics. On the other hand, it appears impossible to explain plate tectonics with the Schrödinger equation.

There is an important distinction between the ontological and epistemological reductions. The philosophical issue of ontological reduction remains an open question. Methodological reduction appears to be false. It is possible that a human being is a quantum mechanical system, but human behaviour is not examined in terms of principles of physics. In a similar manner, even though the Earth is a quantum mechanical system, it is preferable to study it in the context of continuum mechanics. Scientific methods should be considered in the plural. There are overlaps, but each area of science has its own techniques, which are not reducible to those in another branch.

Remark 7.1. It is said that the secondary theory, such as chemistry, reduces to the primary theory, such as physics. It is also common to reverse these terms, even though the same relation is being expressed, such as quantum mechanics reduces to classical mechanics. Usually, the context clarifies the meaning.

In a similar manner, physicists consider isotropy as a special case, since it is a symmetry that results in properties being the same in all directions. Mathematicians, on the other hand, view it as the most general case, since its symmetry group contains all orthogonal transformations. In other words, symmetry groups of all other classes are contained within the group of isotropy.

The opposite of reduction is emergence. A property, P, is emergent, relative to a basic theory, T, if it cannot be predicted on the basis of T or derived from T, nor can T explain any aspect of P.

Example 7.1. Colour of atoms
Colour—as a reflection of light—is an emergent property. An atom alone cannot exhibit such a colour; it is a result of reflection of light from a lattice

formed by a multitude of such atoms. Also, we cannot predict the colour from an assembly of atoms until we know the pattern of their arrangement.

Bunge (2003) is a champion of emergence, claiming that new properties and laws come into existence at different levels of reality. In relation to continuum mechanics, Bunge (1967, pp. 143-144) asks us to consider

> [...] laws in which the stress tensor occurs; since this property is not hereditary (it does not hold for atoms), such laws are characteristic of bulk matter even though we hope to be able to show that they emerge from microlaws: they express patterns that do not exist at the microlevel, just as the latter is characterized by laws of its own.

In other words, the concept of the stress tensor—arguably the most essential entity of continuum mechanics—originates at the macroscopic level, which is the realm of that theory. The stress tensor is a property of the continuum that does not have a counterpart in an atomic structure. This fact suggests that continuum mechanics cannot be reduced to particle mechanics, since— in accordance with its definition in terms of infinitesimal limits—the stress tensor cannot exist in a discrete system, because it requires continuity.

Above, we consider reduction and emergence as ontological issues in an attempt to understand the essence of things. We could also approach the issues from an epistemological viewpoint, and be concerned with procedures and methods for investigating the world. In the seismological context of ray theory, emergence is discussed by Batterman (2002, Section 2.4 and Chapter 8).

Mario Bunge (1967) suggests that—epistemologically—continuum mechanics should be our first theory, since we initially experience the world as consisting of bulk matter. Other theories, such as particle mechanics can be obtained as a discretization of a continuum. This is akin to the integers being a subset of the reals, and—if we start with the latter—we can extract the former.

A pair of heuristic analogies might provide further insight. A dune of fine sand might—at first sight—seem continuous, but since it is composed of discrete grains, it cannot generate a continuum. A sound is perceived as continuous, if the staccato is played with a sufficient speed, even though each note is played individually. This acoustic property was used by the composer György Ligeti in his piece entitled *Continuum*.

It is important to emphasize a distinction between ontological and epistemological reductions. One reduction may be true but not the other. To say that a scientific discipline reduces ontologically to another is to refer to their view of the physical world. This reduction is more modest than the epistemological one, since it allows that—while accepting a common view of the physical world—we might be unable to perform the appropriate methodological reductions.

To say that a discipline reduces epistemologically is to refer to its methods. It claims that—methodologically—we should be able to reduce a given theory to a more fundamental one. The epistemological reduction is rare, in spite of appearances, as illustrated in the preliminary remarks of Chapter 4. Therein, Newton's theory of gravitation, but not Coulomb's law, can be reduced to Newton's mechanics.

7.2   On high-frequency approximation

7.2.1   *Ray theory as approximation*

Within seismology, the relation between wave and ray theories is commonly viewed as analogous to the relationship between quantum and classical mechanics, except that the role of Planck's constant is played by frequency. As frequency tends to infinity, one could say that wave theory approaches ray theory in a manner similar to wave optics approaching geometrical optics (Batterman, 2002, Chapter 6).[1]

However, there are subtleties that appear upon further examination. Ontologically, geometrical optics takes the ray to be its primitive concept (Bunge, 1967, Section 4.2.2). Hence, rays of geometrical optics cannot be derived from another entity. In seismology, rays are derived, but not as a limiting case of another entity. As discussed by Bos and Slawinski (2010), rays are contained in the hyperbolic partial differential equations. Rays are trajectories of propagation of initial conditions,[2] regardless of frequency, which does not explicitly appear in these differential equations. Since rays are not approximations, there is no hierarchy between waves and rays. As shown by Bos and Slawinski (2010), both rays and waves are entailed solely by the equations of motion. Neither of them is an approximation within the mathematical realm, and both can serve as analogies for physical phenomena.

---

[1] Readers interested in relations between the wave and ray theories and between the ray theory and both classical and quantum mechanics might also refer to Batterman (2002, Chapter 6).

[2] *see also*: Bóna and Slawinski (2015, Section 3.4.1)

7.2.2   *Ray theory and banana-doughnut*

In seismological literature, there are many comparisons between the so-called finite-frequency formulation that refers to the insightful work of Dahlen *et al.* (2000) and the so-called infinite-frequency formulation that refers to ray theory. As mentioned in Section 7.2.1, there are issues to consider in examining ray theory in terms of frequency.

Fig. 7.1

Yet, be that as it may, in the aforementioned comparisons, playfully illustrated in Figure 7.1, the accepted belief is that the former, which is commonly referred to as banana-doughnut, is a generalization of the latter, which is viewed as its approximation. For example, the title of Maceira *et al.* (2015) is *On the validation of seismic imaging methods: Finite frequency or ray theory?* In the abstract, the authors describe their work in the following manner.

> We investigate the merits of the more recently developed finite-frequency approach to tomography against the more traditional and approximate ray theoretical approach [...] Based on statistical analyses on *S* wave phase delay measurements, finite frequency shows an improvement over ray theory. Random sampling using cross-correlation values identifies regions where synthetic seismograms computed with ray theory and finite-frequency models differ the most.

However, as proven by Bos and Slawinski (2011, 2013), the finite-frequency formulation is—by invoking standard mathematical properties of crosscorrelation—entailed by the properties contained within ray theory. This formulation allows seismologists to extract information from seismic data that might not be possible to obtain with other applications of ray theory. Hence, it is understandable that such a result—without a concurrent examination of theoretical foundations—fosters the perception of a generalization.

This misperception is, at least partially, explained by the quote of Peirce (c.1896/1955), discussed in Section 1.1, in which he refers to people who know science chiefly by its results. Herein, a method formulated within a theory can be neither its epistemological nor ontological extension, since it is a part of that theory. Notably, as a part of ray theory, banana-doughnut inherits the conceptual and methodological wealth of that theory, which can be used in its applications.

7.3   Relations among models

Seismology offers an intertheory relation among models that represent the Earth. There is an infinity of such models. For instance, a combination of Hookean solids[3] and Stokesian fluids[4] result in the Kelvin-Voigt model[5] and Maxwell model,[6] which differ by the arrangement of this fluid and solid; in the former, they are arranged in parallel and, in the latter, in series.[7] Also, such combinations contain infinite patterns of partial ordering of these arrangements. Moreover, there are many other possible models beside Hookean solids and Stokesian fluids.

Once a model within the continuum is chosen, further specifications must be selected. For a Hookean solid, these specifications are the values of mass density and of elasticity parameters that quantify the elasticity tensor. We refer to the space of these tensors as the space of Hookean solids. Such a space is discussed in detail by Rychlewski (1985), Forte and Vianello (1996), Bóna *et al.* (2008), Danek *et al.* (2013), and in references therein. This space includes infinitely many tensors that satisfy required conditions of a Hookean solid.

---

[3] *see also* : Slawinski (2016, Section 3.2)

[4] *see also* : Slawinski (2016, Section 3.4.2)

[5] *see also* : Slawinski (2016, Section 3.4.3)

[6] *see also* : Slawinski (2016, Section 3.4.4)

[7] Readers interested in descriptions of the Maxwell and the Kelvin-Voigt model models, including the propagation of waves within them, might refer to Roberts (1994, Sections 4.3.1 and 4.3.2), respectively.

7.4   Geology and geophysics

Distinctions between geology and geophysics illustrate the richness and complexity of intertheory relations, if we view them as separate theories, or intratheory relations, if we view both of them as part of a theory of geosciences. A lacunary appreciation of this complexity leads to misguided expectations among representatives of various subjects in geosciences.[8]

Be that as it may, geology is an observational science, whose central concepts are empirical. Geophysics relies on remote sensing, where inferences from observations are mediated by theoretical formulations.

Geology and geophysics associate different meanings with the term understanding. The geological meaning relies on relations among physical objects, which is an ontological emphasis. The geophysical meaning relies on relations between abstract entities and the inference of properties of physical objects, which is an epistemological emphasis.

Oreskes and Doel (2002, p. 544), referring to geophysical and geological approaches, write

> In these recurring debates, a familiar pattern emerges. Geologists argued from qualitative and phenomenological evidence, geophysicists from quantitative and theoretical evidence. Both sides affirmed the superiority of their methods and denied the claim of the other. Geophysicists argued for the greater rigor of mathematical analysis and dismissed empirical counterarguments; geologists defended the accuracy of their observations and frequently dismissed theoretical claims that challenged their conclusions.

However, as discussed by Anduaga (2016, p. 172), many geophysicists, including Jeffreys (1931), exhibit—in contrast to the above passage of Oreskes and Doel (2002)—a propensity for applying an inductive reasoning, with

> the importance of generalizing from sense data to descriptions of nature, rather than logically deducing observations from laws or hypotheses.

Ernest Rutherford declared that all science is either physics or stamp collecting (Birks, 1962). A charitable reading of this statement suggests that some sciences are explanatory by appealing to a theoretical apparatus, while others are descriptive. However, the distinction between explanation

---

[8]Readers interested in a clarification of misguided expectations with respect to seismology might refer to Slawinski (2014).

and description is not sharp, since every science contains both, even though in different proportions. Rutherford insinuates that being descriptive is trivial. However, as discussed on page 8, collecting data is a serious challenge and, to bear fruit, it requires insight and intuition. Such was the case of collecting and correlating geological data on different continents, which served as evidence for plate tectonics.

Furthermore, in the above passage, Anduaga (2016) refers to the process of inductive inference from empirical data, as exemplified by Jeffreys (1931), not to their cataloguing, as suggested by Rutherford's phrase. Hence, as commented in footnote 4, on page 57, seismology might be described as a hypotheticoinferential—not a hypotheticodeductive—science, where its hypotheses consist of abstract models, whose validity is inferred from seismic data. As proposed by Jeffreys (1931), such a validity might be examined in the context of Bayesian probability (Danek and Slawinski, 2012; Tarantola, 2006).[9]

There is a history of interactions between geology and geophysics. Percy William Bridgman, whose philosophical views are mentioned in Section 2.1.2.1, cofounded the Committee on Experimental Geology and Geophysics. As discussed by Anduaga (2016, p. 170–171), the purpose of this committee was

> to reconcile the views of geology and experimental geophysics. [...] the 1932 inauguration at Harvard of a seismological station and of a programme of physical measurements related to Bridgman's research on the field of high pressure amply demonstrated how beneficial—yet also how arduous—reconciling both visions could be.

Such a reconciliation requires an understanding of distinct methodologies of these fields and the recognition of different purposes—at least, *sensu stricto*—for their inferences, even though, as indicated by their prefix, both geology and geophysics share common interests, *sensu lato*.

In view of these common interests, inferences of a seismological theory must not conflict with geological concepts. Consequently, for a seismological theory, it is not enough to be empirically adequate. Among empirically adequate models, a seismologist must select the ones that are consistent with geological data. This necessity imposes restrictions upon solutions of inverse problems. The process of reconciling—broadly speaking—geology and seismology augments the probability of seismological theories succeeding in

---

[9] *see also*: Slawinski (2016, Section 9.3)

giving a true account of the physical world, which is an issue of scientific realism, addressed in Section 2.2.1. According to Alan Musgrave (2006-2007),

> [t]ruth is stronger than empirical adequacy. A true theory will be empirically adequate, but an empirically adequate theory need not be true.

An example of a case in which observable predictions support a theory but do not guarantee its correctness is discussed on page 57.

Also, the relation between geology and geophysics is affected by personal and social values, discussed in Section 2.1.6. Furthermore, such values can affect the relations within geophysics itself. For instance, an administrative choice of geophysics being a part of the physics department or the geoscience department might result in distinct approaches to its foundations and different perceptions of its role as a science. As a consequence of a distinct scientific ethos underlying either department or for a multitude of other reasons, discussions between geophysicists, even though formally belonging to the same discipline, might exhibit difficulties brought to our attention by Charles Percy Snow (1998) in his lecture on the two cultures, in which he refers to the sciences and the humanities.

7.5 Historical relations

Among intratheory relations, there are historical connections that link the present form of a theory to its earlier versions. The appearance of new or modified scientific theories requires an evaluation of their status with respect to their predecessors. Such an evaluation, beyond its obvious importance for the historians of science, is necessary for a fruitful development of theories.

We consider the present form of geophysics as an explanatory science. In antiquity, however, information was collected about times and sizes of tides and about occurrences of earthquakes. Such data collecting is typical for empirical sciences. The explanatory scope of geophysics developed only with sophistication of other branches of physics.

William Gilbert determined that the Earth itself is magnetic, thanks to his researches on magnetism and the observation that compasses point north. Newton was able to explain the tides, thanks to his work in mechanics and the theory of gravitation.

Geophysics develops as an explanatory theory by incorporating advances of physics and mathematics. Also, advances in numerical methods, for data processing and modelling, as well as in technology, for data acquisition and storage, motivate and allow geophysicists to postulate detailed theories to be validated by observations. One might argue that—at least partially—geophysics is data-driven.

7.6   Seismology as field theory

Continuum mechanics underlying the theory of seismology is expressed in the structure of a classical field theory. That means that the effects are local, propagation of effects is continuous and the medium of propagation is a continuum. Furthermore, the speed of propagation is finite, as required by the wave equation.[10]

To emphasize these properties, the general principles of continuum mechanics, discussed in Section 4.2, are commonly referred to as field equations. This terminology is inherited by seismology, since its general principles are field equations of continuum mechanics.

When Michael Faraday and James Clerk Maxwell were creating the theory we now call classical electrodynamics, they faced an interesting situation. What is the status of the field? Is it physically real, like a charged body, or just a useful mathematical device? Conservation of energy and the finite propagation speed led to acceptance of the former. Maxwell (1891/1954, Vol. II, p. 492), complaining about action-at-a-distance theories, remarks,

> [n]ow we are unable to conceive of propagation in time, except as the flight of a material substance through space, or as the propagation of a condition of motion or stress in a medium already existing in space [...] But in all of these theories the question naturally occurs: If something is transmitted from one particle to another at a distance, what is its condition after it has left the one particle and before it has reached the other?

In another place, Maxwell (1890/2012, Vol. I, p. 564) writes.

> [i]n speaking of the energy of the field, however, I wish to be understood literally. All energy is the same as mechanical energy, whether it exists in the form of motion, or in that of elasticity, or in any other form. The energy in electromagnetic phenomena is mechanical energy. The only question is, Where does it

---

[10]*see also*: Slawinski (2015, Section 6.5.4)

reside? On the old theories it resides in the electrified bodies, conducting circuits, and magnets, in the form of an unknown quality called potential energy, or the power of producing certain effects at a distance. On our theory it resides in the electromagnetic field, in the space surrounding the electrified and magnetic bodies, as well as in those bodies themselves, and is in two different forms, which may be described without hypothesis as magnetic polarization, or, according to a very probable hypothesis, as the motion and the strain of one and the same medium.

These remarks imply an argument for the physical reality of the electromagnetic field. Richard Feynman (1963, Vol. I, Ch. 10, p. 9) writes

[t]he fact that the electromagnetic field can possess momentum and energy makes that field very real [...]

Certain concepts invoked in the context of electromagnetic theory of the nineteenth century remain almost unchanged in the theory of elasticity used in seismology, such as the finite speed and conservation of energy within a continuum.[11]

It might appear that the argument for the physical reality of the electromagnetic field carries over to continuum mechanics, and hence, to the continuum in seismology. In the case of seismology, however, there is already a medium to transmit interaction within the Earth. The continuum is a model of the terrestrial material. The same is true for continuum mechanics, in general. The continuum is a model of a liquid, a solid, or any other material, where physical phenomena occur.

Thus, the electromagnetic field has a different status from the continuum of seismology, even though both are media for the transmission of causal processes. The latter has a distinct material counterpart, while the former does not.

In general, the field is either a fiction, introduced for calculation purposes, or a physical entity in its own right. The continuum of seismology is a fiction, whose purpose is to allow calculations that permit seismologists to make inferences about its physical counterpart, which is the terrestrial material.[12]

---

[11]Readers interested in the classical theory of fields and their acquisition of physical reality might refer to Landau and Lifshitz (1975).

[12]Readers interested in the concept of field in the history of physics might refer to Hesse (2005/1961).

Closing remarks

Any seismological inference is an inverse problem, at least *sensu lato*, since the motivation of this science is to learn about properties of the subsurface from available data. Few seismological inferences, however, are achieved by solving an inverse problem *sensu stricto*.

An example of such a solution might be a coordinate-free inference of both the symmetry class and orientation of any Hookean solid expressed as an arbitrary elasticity tensor (Bóna *et al.*, 2007). If that tensor, however, is contaminated by errors—as expected, in view of its experimental or observational origins—this inverse-problem solution needs to be supported by intermediate forward solutions (Danek *et al.*, 2015).

Most seismological inferences are achieved by indirect solutions of inverse problems. Commonly, a specific model within a Hookean framework is postulated and the solution of a forward problem is derived, which consists of hypothetical observables to be compared with measurements. The process is repeated until a satisfactory agreement between the predicted observables and obtained measurements is achieved. Numerical methods and computing facilities allow for this iterative process, which—in its essence— is based on a forward solution.

As discussed in Section 1.4.2, methodologically, the logic of this approach is akin to conjectures proposed by Popper (1959, 1963), except that the model refutations are not as straightforward as he expected. The plethora of available parameter values prevents a simple refutation, as described, in Lakatos's account, by a protective belt.

Afterword

Unfortunately, difficulties never come alone, and in realistic, complex models, a combination of all these methods and more may be needed. Apart from numerical methods, such as finite-difference techniques which are extremely expensive for realistic, complex media at body-wave frequencies, no such comprehensive method has been developed. But enough for today— that is for tomorrow!

Chris Chapman (2004)

Broadly speaking and depending on the chosen emphasis, the theory of seismology can be classified within the natural sciences, in particular, the geosciences, as well as within applied mathematics, mathematical physics and continuum mechanics. The theory of seismology can either precede or follow observations. An experiment can be set up to support a theoretical prediction or a theory can be formulated to retrodict an observation.

To examine such a theory, it is necessary to inquire into relations between abstract entities and observations, and—within the theory—into issues such as reduction, emergence, approximation, generalization and the hierarchy of concepts.

For instance, the formulation of George Backus (1962) is arguably an emergent entity. Therein, a series of parallel layers appear as a single medium with its own properties, provided it is examined through a wave whose wavelength is much greater than the thicknesses of individual layers. One could argue that it is only partially emergent, since—according to the epistemological definition—emergent properties are not predictable from properties of constituents, and, herein, the properties of the medium are derived from the properties of individual layers. However, Mario Bunge (2003, p. 21) discusses the concept of ontological, not epistemological, emergence, which,

> contrary to a widespread opinion, [...] has nothing to do with
> the possibility or impossibility of explaining qualitative novelty.
> Hence, it is mistaken to define an emergent property as a feature
> of a whole that cannot be explained in terms of properties of
> its parts. Emergence is often intriguing but not mysterious:
> explained emergence is still emergence.

Be that as it may, the inferences obtained from seismological studies
might appear in both qualitative and quantitive disagreement with detailed
laboratory or geological investigations, unless the intertheory and intratheory relations are understood.

To understand these relations, let us examine Figure A.1. This figure
contains two realms: physical and abstract. The relation between the physical realm and the abstract realm is varied and complex. It is important
to find structures in one that are similar to structures in the other. Thus,
as discussed in Section 3.1, the Earth is similar, in certain respects, to a
Hookean solid, which in turn is similar to $\mathbb{R}^3$. Discovering these similarities
and elaborating their details is the essence of theoretical seismology.

In Figure A.1, the physical realm contains such object as the Earth and
the Solar system. The abstract realm contains the theories of seismology,
continuum mechanics and mathematics, which are distinct theories related
by the concept of continuum. The analytic continuum of mathematics,
discussed in Section 3.1, serves as a model for the mechanical continuum,
which—in turn—serves as a model for the physical continuum of seismology, to which we also refer as the seismological continuum, as discussed in
Section 3.5.

The abstract realm also contains such structures as musical notation,
whose physical counterparts are sounds, which are mechanical vibrations
of the air. This relation is similar to the case of Hookean solids, whose
counterparts are physical materials. Also, similar to notes, which prescribe
duration and pitch but are not constrained to any particular instrument,
Hookean solids, which prescribe density and mechanical properties of a
solid, are not constrained to any particular material.

Seismology and continuum mechanics belong to an ensemble of fictions
that also contains Sherlock Holmes, but—as illustrated by the comic strip
in Figure 3.2 on page 55—not π. Mathematics, which includes π, is distinct
from both seismology and continuum mechanics, which use mathematical
entities, such as the analytic continuum and relations therein, as analogies

of physical concepts. Continuum mechanics is distinct from seismology, which uses its concepts as models of the Earth. As remarked on page 77, the theory of seismology relies on the structure of continuum mechanics. It is not, however, just an application of continuum mechanics to terrestrial materials. This theory contains concepts and principles not included in continuum mechanics.

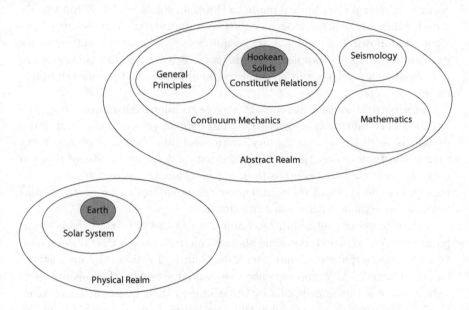

Fig. A.1   Physical objects, such as the Earth, belong to a different realm than abstract entities, such as Hookean solids.

Continuum mechanics contains the mechanical continuum, as discussed in Section 3.1. It contains the primitive concepts and first principles, as discussed in Section 3.2. It also contains its two distinct parts, as discussed in Section 4.1; they are the general principles of Section 4.2 and constitutive relations of Section 4.3.

By establishing continuum mechanics as its framework, theoretical seismology becomes an idealization, which accepts the falsity of certain postulated assumptions, and fictions that are entailed.

Among such fictions of seismology are P and S waves, as discussed in Section 1.2. This is a consequence of the fact that the continuum model determines the interpretation of seismic measurements, since the specific theory must be invoked to mediate between data and information derived from these data. Such a statement might, at least at first sight, appear inappropriate to an empiricist. However, for a science that relies on remote sensing, such an approach is necessary, and its justification is supported by a conviction of satisfactory results.

The theory of seismology is in the abstract realm but engages both realms. It relates the physical realm to Hookean solids, and—within the abstract realm—it relates Hookean solids to a mathematical structure. Consider the following example. A set of numbers is an abstract entity, as are its members. A set of apples is an abstract entity, even though apples are not. Seismology relates the physical to the abstract, and such a relation is an abstract entity.

As a natural science, seismology strives to infer information about the Earth. This information consists of mechanical properties of terrestrial materials, such as density, rigidity, compressibility, existence of symmetry planes and their orientations. This information does not consist of types of rocks or their chemical composition, nor can they be derived from it, even though certain types of rocks and their composition can be postulated as being in agreement with seismic information.

Upon invoking continuum mechanics, we abandon certain levels of explanation. We explain large-scale phenomena in a manner that is more convenient, and perhaps even more reliable, than by invoking the microstructure of matter. As a consequence, we cannot describe the seismological behaviour of solids and liquids by appealing to their common microstructure. In other words, we abandon the possibility of accounting for their behaviours in terms of common underlying features. There is no possibility—within continuum mechanics—of a unified account of such behaviours, and hence, we must invoke constitutive relations.

In the context of the geosciences, the concept of continuum allows seismologists a description that—by its very nature—abandons direct information about grains, crystals, laminations, and other discrete features that might be essential for geological studies. Obtaining such information is not the purpose of seismology, even though it might be achieved by—and with limitations of—indirect processes, akin to inferring information about an individual object from the average of a sample containing a large number of objects.

With all its limitations, and due to resulting immediacy of ontological insights and epistemological conveniences, the theory of seismology allows us to infer coherent information about the Earth, together with satisfactory predictions of terrestrial phenomena. As an *envoi*, let us consider that scientific interests in seismology can be categorized as contemplative or manipulative, where, in the words of Hutchinson (1953),

> [t]he only real values which we seem to experience are, in this
> categorization, contemplative ones. [...] In so far as applied
> science defends our culture and raises our standard of living, it
> is good only if the culture contains contemplative values [...]

Bibliography

Abraham, R. H. and Shaw, C. D. (1992). *Dynamics: The geometry of behavior* (Addison-Wesley).

Achenbach, J. D. (1973). *Wave propagation in elastic solids* (North-Holland Publishing Co.).

Aki, K. and Richards, P. G. (2002). *Quantitative seismology* (University Science Books).

Anduaga, A. (2016). *Geophysics, realism, and industry: How commercial interests shaped geophysical conceptions, 1900–1960* (Oxford University Press).

Aoki, H., Syono, Y. and Hemley, R. (eds.) (2000). *Physics meets mineralogy: Condensed-matter physics in geosciences* (Cambridge University Press).

Arnold, V. I. (1989). *Mathematical methods of classical mechanics*, 2nd edn. (Springer-Verlag).

Arnold, V. I. (1992). *Catastrophe theory*, 3rd edn. (Springer-Verlag).

Babich, V. M. and Buldyrev, V. S. (2009). *Asymptotic methods in short-wavelength diffraction theory* (Alpha Science).

Backus, G. (1962). Long-wave elastic anisotropy produced by horizontal layering, *JGR* **67**, pp. 4427–4440.

Batterman, R. (2002). *The Devil in the details: Asymptotic reasoning in explanation, reduction, and emergence* (Oxford University Press).

Berry, M. V. and Upstill, C. (1980). *Catastrophe optics: Morphologies of cautics and their diffraction patterns* (North-Holland Publishing Co.).

Birks, J. B. (ed.) (1962). *Rutherford at Manchester* (Heywood).

Bogan, J. and Woodward, J. (1945). Saving the phenomena, *Philosophical Review* **97**, 3, pp. 303–352.

Bóna, A., Bucataru, I. and Slawinski, M. A. (2004a). Characterization of elasticity-tensor symmetries using SU(2), *Journal of Elasticity* **75**, 3, pp. 267–289.

Bóna, A., Bucataru, I. and Slawinski, M. A. (2004b). Material symmetries of elasticity tensor, *The Quarterly Journal of Mechanics and Applied Mathematics* **57**, 4, pp. 583–598.

Bóna, A., Bucataru, I. and Slawinski, M. A. (2007). Coordinate-free characterization of elasticity tensor, *Journal of Elasticity* **87**, 2-3, pp. 109–132.

Bóna, A., Bucataru, I. and Slawinski, M. A. (2008). Space of SO(3)-orbits of elasticity tensors, *Archives of Mechanics* **60**, 2, pp. 121–136.

Bóna, A. and Slawinski, M. A. (2015). *Wavefronts and rays as characteristics and asymptotics*, 2nd edn. (World Scientific).

Bos, L., Dalton, D. R., Slawinski, M. A. and Stanoev, T. (2016). On Backus average for generally anisotropic layers, *Journal of Elasticity* **DOI 10.1007/s10659-016-9608-z**.

Bos, L. and Slawinski, M. A. (2010). Elastodynamic equations: Characteristics, wavefronts and rays, *The Quarterly Journal of Mechanics and Applied Mathematics* **63**, 1, pp. 23–37.

Bos, L. and Slawinski, M. A. (2011). Proof of validity of first-order seismic traveltime estimates, *International Journal on Geomathematics* **2**, 2, pp. 255–263.

Bos, L. and Slawinski, M. A. (2013). On the relationship between ray theory and the banana-doughnut formulation, *International Journal on Geomathematics* **4**, 1, pp. 55–65.

Box, G. E. P. and Draper, N. R. (1987). *Empirical model-building and response surfaces* (Wiley).

Boyd, R., Gasper, P. and Trout, J. D. (eds.) (1991). *The Philosophy of science* (MIT Press).

Brown, J. R. (2008). *Philosophy of mathematics: A contemporary introduction to the world of proofs and pictures*, 2nd edn. (Routledge).

Brush, S. G. (1980). Discovery of the Earth's core, *American Journal of Physics* **48**, 9, pp. 705–724.

Brush, S. G. (2009). *A history of modern planetary physics: Nebulous Earth*, Vol. 1 (Cambridge University Press).

Bunge, M. (1967). *Foundations of physics* (Springer-Verlag).

Bunge, M. (1998). *Philosophy of science, Vol. I: From problem to theory (Revised edition)* (Transaction Publishers).

Bunge, M. (2003). *Emergence and convergence: Qualitative novelty and the unity of knowledge* (University of Toronto Press).

Cannell, D. M. (2001). *George Green, mathematician and physicist: The background to his life and work* (SIAM).

Carter, M. and van Brunt, B. (2000). *The Lebesgue-Stieltjes integral* (Springer).

Cartwright, N. (1994). *How the laws of physics lie* (Oxford University Press).

Chapman, A. (2005). *England's Leonardo: Robert Hooke and the seventeenth-century scientific revolution* (IOP Publishing Ltd).

Chapman, C. H. (2004). *Fundamentals of seismic wave propagation* (Cambridge University Press).

Close, F. (2009). *Nothing: A very short introduction* (Oxford University Press).

Colyvan, M. (2001). *The indispensability of mathematics* (Oxford University Press).

Cosserat, E. M. P. and Cosserat, F. (1909). *Théorie des corps déformables* (Hermann et fils).

Curd, M. and Cover, J. (eds.) (1998). *Philosophy of science: The central issues* (Norton).

Dahlen, F. A., Hung, S.-H. and Nolet, G. (2000). Fréchet kernels for finite-frequency traveltimes — I. Theory, *Geophys. J. Int.* **141**, pp. 157–174.

Dahlen, F. A. and Tromp, J. (1998). *Theoretical global seismology* (Princeton University Press).

Danek, T., Kochetov, M. and Slawinski, M. A. (2013). Uncertainty analysis of effective elasticity tensors using quaternion-based global optimization and Monte-Carlo method, *The Quarterly Journal of Mechanics and Applied Mathematics* **66**, 2, pp. 253–272.

Danek, T., Kochetov, M. and Slawinski, M. A. (2015). Effective elasticity tensors in context of random errors, *Journal of Elasticity* **121**, 1, pp. 55–67.

Danek, T. and Slawinski, M. A. (2012). Bayesian inversion of VSP traveltimes for linear inhomogeneity and elliptical anisotropy, *Geophysics* **77**, 6, pp. 239–243.

Davison, C. (1927/2014). *The founders of seismology* (Cambridge University Press).

Doxiadis, A. and Papadimitriou, C. H. (2009). *Logicomix* (Bloomsbury).

Duhem, P. (1914). *La Théorie physique: Son objet, sa structure*, 2nd edn. (Marcel Rivière & Cie.).

Duhem, P. (1991). *The aim and structure of physical theory* (Princeton University Press).

Dyson, F. (2001). *George Green, mathematician and physicist: The background to his life and work*, chap. Homage to George Green: How physics looked in the nineteen-forties (SIAM).

Dyson, F. (2014). The case for blunders, *The New York Review of Books* **61**, 4, pp. 4–8.

Dziewoński, A. M. and Anderson, D. L. (1981). Preliminary reference earth model, *Physics of the Earth and Planetary Interiors* **25**, 4, pp. 297–356.

Einstein, A. (1949). *Albert Einstein: philosopher-scientist*, chap. Autobiographical notes (Open Court).

Einstein, A. and Infeld, L. (1938). *Evolution of physics* (Simon and Shuster).

Epstein, M. (2010). *The geometrical language of continuum mechanics* (Cambridge University Press).

Feyerabend, P. K. (1975). *Against method* (New Left Books).

Feynman, R. P. (1963). *The Feynman lectures in physics (3 Vols.)* (Addison-Wesley).

Feynman, R. P. (1985/2006). *QED: The strange theory of light and matter* (Princeton University Press).

Forest, S. (2006). *Milieux continus généralisés et matériaux hétérogènes* (Mines Paris).

Forte, S. and Vianello, M. (1996). Symmetry classes for elasticity tensors, *Journal of Elasticity* **43**, 2, pp. 81—108.

Fowler, A. C. (1997). *Mathematical models in the applied sciences* (Cambridge University Press).

Giere, R. (2004). How models are used to represent reality, *Philosophy of Science* **71**, pp. 742–752.

Goldstein, H. (1980). *Classical mechanics*, 2nd edn. (Addison-Wesley).

Gowers, T. (ed.) (2008). *Princeton companion to mathematics* (Princeton University Press).

Graney, C. M. (2015). *Setting aside all authority: Giovanni Battista Riccioli and the science against Copernicus in the age of Galileo* (Notre Dame Press).

Guidoboni, E. and Poirier, J.-P. (2004). *Quand la Terre tremblait* (Odile Jacob).

Hadamard, J. (1932). *Le problème de Cauchy et les équations aux dérivées partielles linéaires et hyperboliques* (Édition Jacques Gabay).

Hempel, C. (1970). *Aspects of scientific explanation and other essays in the philosophy of science* (Free Press).

Herman, B. (1945). Some theorems of the theory of anisotropic media, *Comptes Rendus (Doklady) de l'Académie des Sciences de l'URSS* **48**, 2, pp. 89–92.

Hesse, M. B. (2005/1961). *Forces and fields: The concept of action at a distance in the history of physics* (Dover).

Hildebrandt, S. and Tromba, A. (1996). *The parsimonious universe: Shape and form in the natural world* (Springer-Verlag).

Hough, S. E. (2007). *Richter scale: Measure of an earthquake, measure of a man* (Princeton University Press).

Hume, D. (1739/2007). *A treatise of human nature* (Oxford University Press).

Hutchinson, G. E. (1953). *The itinerant ivory tower: Scientific and literary essays* (Yale University Press).

Jeffreys, H. (1931). *Scientific inference* (Cambridge University Press).

Kac, M. (1966). Can one hear the shape of a drum? Part II, *American Mathematical Monthly* **73**, 4, pp. 1–23.

Kennett, B. L. N. and Bunge, H.-P. (2008). *Geophysical continua: Deformation in the Earth's interior* (Cambridge University Press).

Kitcher, P. (1983). *The nature of mathematical knowledge* (Oxford University Press).

Kleinhans, M. G., Buskes, C. J. J. and de Regt, H. W. (2005). Terra incognita: Explanation and reduction in earth science. *International Studies in Philosophy of Science* **19**, 3, pp. 289–317.

Koks, D. (2006). *Exploration in mathematical physics: The concepts behind an eloquent language* (Springer).

Kot, M. (2014). *A first course in the calculus of variations* (AMS).

Koyré, A. (1968). *Newtonian studies* (University of Chicago Press).

Kuhn, T. S. (1996). *The structure of scientific revolutions*, 3rd edn. (The University of Chicago Press).

Lakatos, I. (1970). The methodology of scientific research programmes, in Lakatos and Musgrave (eds.), *Criticism and the growth of knowledge* (Cambridge University Press).

Lanczos, C. (1986/1970). *The variational principles of mechanics*, 4th edn. (Dover).

Landa, E. and Treitel, S. (2016). Seismic inversion: What it is, and what it is not, *The Leading Edge* **35**, 3, pp. 277–279.

Landau, L. D. and Lifshitz, E. M. (1975). *The classical theory of fields* (Pergamon).

Laplace, P.-S. (1812). *A philosophical essay on probabilities.*

Lefebvre, M. (2012). *Algorithmes sur GPU pour la simulation numérique en mécanique des fluides*, Ph.D. thesis, Université Paris 13.

Leng, M. (2010). *Mathematics and reality* (Oxford University Press).

Lipton, P. (2004). *Inference to the best explanation*, 2nd edn. (Routledge).

Love, A. E. H. (1944). *A treatise on the mathematical theory of elasticity* (Dover).

Maceira, M., Larmat, C., Porritt, R. W., Higdon, D. M., Rowe, C. A. and Allen, R. M. (2015). On the validation of seismic imaging methods: Finite frequency or ray theory? *Geophysical Research Letters* **42**, 2, pp. 323–330.

Malvern, L. E. (1969). *Introduction to the mechanics of a continuous medium* (Prentice-Hall).

Marsden, J. E. and Hughes, T. J. R. (1983). *Mathematical foundations of elasticity* (Dover).

Maugin, G. A. (2013). *Continuum mechanics through the twentieth century: A concise historical perspective* (Springer).

Maxwell, J. C. (1890/2012). *The scientific papers of James Maxwell* (Dover).

Maxwell, J. C. (1891/1954). *Treatise on electricity and magnetism* (Dover).

McAllister, J. (2007). Model selection and the multiplicity of patterns in empirical data, *Philosophy of Science* **74**, 2, pp. 884–894.

McMullin, E. (1985). Galilean idealization, *Studies in the history and philosophy of science* **16**, 3, pp. 247–273.

Mill, J. S. (1919). *A system of logic, ratiocinative and inductive* (Longmans, Green, and Co.).

Morrison, M. (2015). *Reconstructing reality: Models, mathematics, and simulations* (Oxford University Press).

Müller, W. H. and Weiss, W. (2016). *The state of deformation in Earthlike self-gravitating objects* (Springer).

Musgrave, A. (2006-2007). The 'miracle argument' for scientific realism, *The Rutherford Journal* **2**.

Nola, R. and Sankey, H. (2007). *Theories of scientific method* (McGill-Queen's University Press).

Nye, J. F. (1999). *Natural focusing and fine structure of light: Caustics and wave dislocations* (IOP Publishing Ltd).

Ogden, R. W. (1997/1984). *Non-linear elastic deformations* (Dover).

Oreskes, N. and Doel, R. E. (2002). Physics and chemestry of the Earth, in M. J. Nye (ed.), *The Cambridge history of science*, Vol. 5 (Cambridge University Press), pp. 538–552.

Peirce, C. S. (c.1896/1955). *Philosophical writings of Peirce* (Dover).

Penrose, R. (1997). *The large, the small and the human mind* (Cambridge University Press).

Penrose, R. (2004). *The road to reality: A complete guide to the laws of the Universe* (Random House).

Pestov, L. and Uhlmann, G. (2005). Two dimensional compact simple Riemannian manifolds are boundary distance rigid, *Annals of mathematics* **161**, pp. 1093–1110.

Placanica, A. (1997). *Il filosofo e la catastrofe. Un terremoto del Settecento* (Einaudi).

Poincaré, H. (1968). *La science et l'hypothèse* (Flammarion).

Popper, K. (1959). *The logic of scientific discovery* (Routledge).

Popper, K. (1963). *Conjectures and refutations* (Routledge).

Porteous, I. R. (1994). *Geometric differentiation for the intelligence of curves and surfaces* (Cambridge University Press).

Pour-El, M. and Richards, I. (1983). Noncomputability in analysis and physics: A complete determination of the class of noncomputable linear operators, *Advances in Mathematics* **48**, 1, pp. 44–74.

Richter, C. F. (1935). An instrumental earthquake magnitude scale, *Bulletin of the Seismological Society of America* **25**, 1-2, pp. 1–32.

Rivlin, R. S. (1996). On the principles of equipresence and unification, in G. Barenblatt and D. Joseph (eds.), *Collected papers of R.S. Rivlin*, Vol. 1 (Springer-Verlag).

Roberts, A. J. (1994). *A one-dimensional introduction to continuum mechanics* (World Scientific).

Robinson, E. A. and Treitel, S. (2000). *Geophysical signal analysis* (Society of Exploration Geophysicists).

Rochester, M. and Crossley, D. (1987). Earth's third ocean: The liquid core, *EOS* **68**, 17, pp. 491–492.

Rogister, Y. and Slawinski, M. A. (2005). Analytic solution of ray-tracing equations for a linearly inhomogeneous and elliptically anisotropic, *Geophysics* **70**, 5, pp. D37–D41.

Rychlewski, J. (1985). On Hooke's law, *Prikl. Matem. Mekhan.* **48**, 3, pp. 420–435.

Shapiro, S. (2000). *Thinking about mathematics: The philosophy of mathematics* (Oxford University Press).

Slawinski, M. A. (2014). On Hookean solids in seismology: anisotropy and fractures, *Recorder* **39**, 2, pp. 24–31.

Slawinski, M. A. (2015). *Waves and rays in elastic continua*, 3rd edn. (World Scientific).

Slawinski, M. A. (2016). *Waves and rays in seismology: Answers to unasked questions* (World Scientific).

Slawinski, M. A., Wheaton, C. J. and Powojowski, M. (2004). VSP traveltime inversion for linear inhomogeneity and elliptical anisotropy, *Geophysics* **69**, 2, pp. 373–377.

Slotnick, M. M. (1959). *Lessons of seismic computing* (Society of Exploration Geophysicists).

Smith, D. (2012). *Former people: The final days of the Russian aristocracy* (Ferar, Straus and Giroux).

Smith, G. (2007a). Gaining access: Using seismology to probe the earth's insides (unpublished lecture on seismology), URL `http://web.stanford.edu/dept/cisst/events0506.html`.

Smith, S. (2007b). Continuous bodies, impenetrability, and contact interactions: The view from the applied mathematics of continuum mechanics, *British Journal for the Philosophy of Science* **58**, pp. 503–538.

Smolin, L. (2007). *The trouble with physics: The rise of String Theory, and fall of science, and what comes next* (Mariner Books).

Snow, C. P. (1998). *The two cultures* (Cambridge University Press).

Stebbing, L. S. (1958/1937). *Philosophy and the physicists* (Dover).

Stefanov, P., Uhlmann, G. and Vasy, A. (2017). Local and global boundary rigidity and the geodesic X-ray transform in the normal gauge, *arXiv*.

Suppes, P. (1962). Models of data, in E. Nagel, P. Suppes and A. Tarski (eds.), *Logic, methodology, and philosophy of science*, pp. 252–261.

Tarantola, A. (2006). Popper, Bayes and the inverse problem, *Nature Physics* **2**, pp. 492–494.

Thom, R. (1993). *Prédire n'est pas expliquer* (Flammarion).

Trèves, F. (2006). *Basic linear partial differential equations* (Dover).

Truesdell, C. A. (1966). *Six lectures on modern natural philosophy* (Springer-Verlag).

van Fraassen, B. (2008). *Scientific representation: Paradoxes of perspective* (Oxford University Press).

Verne, J. (1864). *Voyage au centre de la Terre* (Hachette).

Voltaire (1734). *Discours en vers sur l'homme: De la modération en tout*, Vol. 4.

Weinberg, S. (2001). *Facing up: Science and its cultural adversaries* (Harvard University Press).

Weisberg, M. (2013). *Simulation and similarity* (Oxford University Press).

Woodward, J. (2009). Scientific explanation, URL `http://plato.stanford.edu/entries/scientific-explanation/`.

Zee, A. (2016). *Group theory in a nutshell for physicists.* (Princeton University Press).

Subject Index

Abel problem, 111
absolute space, 101
abstract realm, 7, 10, 56, 67, 97, 101, 128, 130
   see also physical realm
action-at-a-distance, 5, 84, 124
adiabatic process, 99, 100
analytic continuum, 50, 52, 54–56, 60, 89, 128
antirealism, 29, 30, 45, 47, 64
   see also realism
applied science, 131
Aristotelian idealization, 72, 73
   see also Galilean idealization
asteroseismology, 1
astrology, xix, 23
astronomy, 20, 30, 39, 44, 46
astrophysics, 8
asymptotic method, xx
asymptotic reasoning, xx
axiom of choice, 53

background radiation, 86
Backus average, 54, 74, 98, 127
balance of
   angular momentum, 78, 81, **81**, 82
   electric charge, 82
   energy, 82
   linear momentum, 74, 78, **81**, 82, 95, 106
   magnetic flux, 82
   mass, 78, **80**, 82

   rotational momentum, 95
banana-doughnut, 119, 120
   see also finite-frequency formulation
Bayesian probability, 122
Bayesian statistics, 17
begging the question, 112
body force, 81, 106, 108
   gravity, 106

catastrophe theory, 62
Cauchy data, 107, 108
Cauchy problem, 108
Cauchy's stress principle, 63
Cauchy-elastic material, 87
   see also simple elastic material
Cauchy-Kovalevskaya theorem, 108
caustic, 62
Chandler wobble, 44
classical mechanics, 46, 66, 77, 116
   continuum mechanics, 58
   Hamiltonian, 46
   Lagrangian, 46
   Newtonian, 46
   quantum mechanics, xx
   ray theory, 118
cognitive value, 41
Committee on Experimental Geology and Geophysics, 122
condensed-matter physics, 57, 76
   continuum mechanics, 57, 58
conservation laws, 78

Name Index

About the Authors and Illustrator

James Robert Brown

J. R. Brown is a Professor of Philosophy at the University of Toronto. His interests include a range of topics in the philosophy of science and mathematics: thought experiments, foundational issues in mathematics and physics, visual reasoning, and issues involving science and society, such as the detrimental role of commercialization in medical research.

He is the author of several books, including: *The Laboratory of the Mind: Thought Experiments in the Natural Science* (Routledge 1991/2010), *Philosophy of Mathematics: An Contemporary Introduction to the World of Proofs and Pictures* (Routledge 1999/2008), *Who Rules in Science: An Opinionated Guide to the Wars* (Harvard 2001).

He has been elected to the Royal Society of Canada, the German Academy of Sciences (Leopoldina), and l'Académie Internationale de Philosophie des Sciences.

Michael A. Slawinski

M. A. Slawinski is a Professor of Seismology at Memorial University in St. John's and the director of The Geomechanics Project, a theoretical-research group—founded in 1998 at the University of Calgary—whose focus is on extending seismological theory within the realm of continuum mechanics and the language of mathematics.

He has long-term research collaborations with Department of Mathematics at Politecnico di Milano, Department of Geosciences at Princeton University, Department of Philosophy at University of Toronto, and Department of Geoinformatics and Applied Computer Science at University of Science and Technology in Kraków.

He appears in The Canadian Encyclopedia as a specialist of the nature of applied mathematics. This is his fourth book with World Scientific.

www.thegeomechanicsproject.com

Roberto Lauciello

Roberto Lauciello (LAU) is a comic artist, children's books illustrator, cartoonist and character designer. He lives and works in Rapallo, Italy, and is virtually accessible from virtually everywhere.

robertolauciello.blogspot.it

Printed in the United States
By Bookmasters